THE FABLE
THE FOSSILS
AND THE FLOOD

David A. Meunier, M.S.

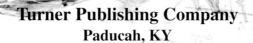

Turner Publishing Company
Paducah, KY

TURNER PUBLISHING COMPANY
412 Broadway • P.O. Box 3101
Paducah, Kentucky 42002-3101
(270) 443-0121

Turner Publishing Company Staff:
Editor: Dayna Spear Williams
Designer: Susan L. Harwood

Library of Congress Catalog Control No: 00-112278
ISBN: 1-56311-662-6

Cover Art taken from: *Dinosaurs by Design*
Published by Master Books, Used by permission
Illustrations used on pages 15, 65 and 71 taken from: *The Young Earth*
Published by Master Books, Used by permission

Printed in the United States of America. Additional copies may be
purchased directly from the publisher. Limited Edition.

TABLE OF CONTENTS

CHAPTER 1

GENESIS IS THE FOUNDATION

This book is an examination of the creation/evolution controversy from a conservative Christian viewpoint. It attempts to show the basic unscientific nature of evolution, and its far reaching influence on history. Although the primary purpose is not evangelistic, it is hoped that by reading this book, Christians will have their faith strengthened and be better able "to give an answer to every man that asketh you a reason of the hope that is in you." (I Peter 3:15).

Dr. Henry Morris, a pioneer creationist and hydraulic engineer, writes, "Evolution is not merely a biological theory of little significance. It is a world view - the world view - diametrically opposing the Christian world view. Therefore, Christians ignore it or compromise with it at great peril!" Real science is characterized by observation, measurement and repeatability. Yet no scientist has really observed evolution taking place. Even if evolution were true, it happens too slowly to be measured and has progressed by a series of non-repeatable steps. Therefore, evolution is beyond the scope of true science, and has not been proven by the scientific method.[1]

Ken Ham, an Australian creationist and former public school teacher, writes, "There are two world views with two totally different belief systems clashing in our society. The real war being waged is a great spiritual war. Sadly, today many Christians fail to win the war because they fail to recognize the nature of the battle. It is my contention that this spiritual conflict is rooted in the issue of origins. Although the thought may sound strange or new to the reader, Biblically and logically this issue is central in the battle for men's souls."[2] Evolution actually resembles religion more than it resembles science. Since no scientist was present to observe the alleged long ages of evolutionary progression, evolution is really a belief system concerning origins.[3] Evolution was embraced mainly for philosophical reasons. People were looking for an explanation of origins that was naturalistic. They did not like the idea of a special creation by God. The implication of creation is that people have a responsibility to the Creator, which is why many sought a naturalistic theory of origins.[4]

Evolution and special creation can be contrasted as competing models. It is not possible to experimentally prove either side since past history is neither observable or reproducible in the laboratory. Dr. Morris wrote, "We can define two models of origins, and then make comparative predictions about what our observations should find if evolution is true, and conversely, what we should find if creation is true. The model that enables us to do the best job of predicting things that we then find to be true on observation is the model most likely to be true although we cannot prove it to be true."[5]

If the evolutionists are correct in their belief that all plants and animals "descended with modification from a single form of life", then it should not be difficult to find these intermediate stages in the fossil record. In fact, millions of fossils have been collected and catalogued with over 250,000 fossil species represented. Since Charles Darwin's famous book, *The Origin of Species*, was published in 1859, evolutionists have searched for these intermediate forms that the evolution model predicts. If this model is correct, then these transitional forms (between invertebrates and fishes, between reptiles and birds, and so on) should be at least as numerous as the terminal forms. At this point, many thousands of these transitional forms should have been collected and clearly identified. That is, if the evolution model is accurate. By contrast, if the creation model is accurate, then each type of plants and animals would be clearly distinguished from all others and would never appear in the fossil record less than fully formed. Dr. Duane Gish, who holds a Ph.D. in biochemistry, summarizes the creation model as, "the abrupt appearance, fully formed, of each created type and the total absence of organisms that could be interpreted as constituting transitional forms between these basic types." Each of these types, similar in morphological design would share a common genetic origin.

The creation model is the one that is actually supported by the facts. These multitudes of transitional forms, or even a few of them, simply do not exist. Darwin was aware of this problem, but hoped that these intermediates would someday be discovered. The absence of these transitional forms has been an embarrassment to evolutionists.[6] As William Jennings Bryan has written, "If it were true that all species came by slow development from one or a few germs every square foot of the earth's surface would teem with evidences of change. If everything changed, we ought to find evidence of it somewhere, but because it is not true, never was true and seemingly cannot be true, they have not found one thing, living or dead, in process of change."[7]

According to Dr. Gish, evolution does not qualify as a scientific theory because it meets none of the required criteria. He writes that, "to qualify as a theory, it must be supported by events, processes, or properties that can be observed and the theory must be useful in predicting the outcome of future natural phenomena or laboratory experiments. An additional limitation usually imposed is that the theory must be capable of falsification, that is, it must be possible to conceive some experiment, the failure of which would disprove the theory." Evolutionists often say that Biblical creation must be rejected as an explanation for origin because it cannot be tested by the experimental method. However, when creationists say the same thing about evolution, the evolutionists attempt to dodge the issue by claiming that the lengths of time necessary for evolution are too great for experimental observation. But by saying this, they are really admitting that evolution is at best a postulate (i.e., a presupposition) which cannot be proven.[8]

When confronted with questions about evolution, such as previously mentioned, many Christians will ask if it really matters. It does matter because the creation/evolution discussion has everything to do with the historical and scientific accuracy of Genesis. The accuracy, or perhaps the reliability, of Genesis is important to Christians because of the fact that Genesis is a foundation for the rest of the Bible. As Ken Ham has written, "The Biblical doctrine of origins, as contained in the book of Genesis, is foundational to all other doctrines of Scripture. Refute or undermine in any way the Biblical doctrine of origins, and the rest of the Bible is compromised."[9]

Dr. Henry Morris writes in *Biblical Creationism* that the Genesis account of creation is the "foundation of all Christian doctrine" and lists ten doctrines based on the biblical doctrine of creation. The first three on his list are the foundations of true Christology, the true gospel, and the true faith. He writes, "Creation is the foundation of true Christology" because of Hebrews 1:2-3 that clearly says that the same Son of God who "made the worlds" also "by himself purged our sins." Dr. Morris makes the point that unless we present Christ the Savior as Christ the Creator, we are guilty of teaching "another Jesus" as referred to in II Corinthians 11:4. Dr. Morris gives Revelation 14:6-7 as a reference, "I saw another angel fly in the midst of heaven, having the everlasting gospel to preach unto them that dwell on the earth,...and worship him that made heaven, and earth, and the sea, and the fountains of water." We find here that creation is presented as a part of the gospel. Hebrews 11:3 is given as a reference to support the idea that creation is the foundation of true faith. In this passage we find that, "through faith we understand that the worlds were framed by the word of God." It is remarkable that this great faith chapter in the Bible begins with a reference to the belief in a special creation. This verse was probably also written to negate the concept of theistic evolution. Theistic evolution is basically the belief that God used previously existing materials to gradually form the present earth and inhabitants over long ages of time.

The next three doctrines listed as having their foundations in creation are those of true evangelism, true missions, and of true Bible teaching. Almost at the beginning of the gospel of John is the declaration that "All things were made by him; and without him was not anything made that was made" (John 1:3). John puts a priority on establishing that Christ the Savior is also Christ the Creator in this great evangelistic book of the Bible. Dr.

Morris also gives the example of Paul who taught polytheistic evolutionists about the "living God, which made heaven, and earth, and the sea, and all things that is therein" (Acts 14:15) on his missionary journey to Lystra. Another example given was that of Paul preaching about the "God that made the world and all things therein" (Acts 17:24). Creation must also be at the foundation of Bible teaching since Christ, when teaching the two disciples on the road to Emmaus began with Genesis as "he expounded unto them in all the Scriptures the things concerning himself" (Luke 24:27).

The last four doctrines that Dr. Morris lists as having their foundation in creation are those of true fellowship, true marriage and family relations, all honest human vocations, as well as the Christian life itself. When Dr. Morris writes about the foundation of true fellowship, he uses Ephesians 3:9 as a reference where it says, "to make all men see what is the fellowship of the mystery, which from the beginning of the world hath been hid in God, who created all things by Jesus Christ." In this passage, the God through whom we have fellowship is clearly identified as the Creator. Matthew 19:4-6 is used as the reference to support the idea that creation is at the foundation of marriage. In this passage, when Jesus refers to marriage, He uses creation as its basis. Dr. Morris writes that "all honorable human occupations" are included in God's command to subdue the earth in Genesis 1:28, so that vocations are a result of creation. In Colossians 3:10, the Bible says, "...and have put on the new man, which is renewed in knowledge after the image of him that created him." In this passage, which refers to Christian growth, we are again brought back to creation.

It is impossible to reconcile evolution with the doctrine of original sin. In Genesis 2:15-17, the Bible says, "And the Lord God took the man (Adam), and put him into the garden of Eden to dress it and to keep it. And the Lord God commanded the man, saying, of every tree of the garden thou mayest freely eat: but of the tree of knowledge of good and evil, thou shalt not eat of it: for in the day that thou eatest thereof thou shalt surely die." In the second chapter of Genesis we find that Adam was given a warning that willful disobedience to God's command carried the penalty of death. Then we find the account of Adam's sin, God's curse on the ground, and the expulsion of Adam and Eve from the garden in Genesis 3. However, the full consequences of Adam's sin are not explained until later in the Bible.[10]

We find in Romans 5:12, "Wherefore, as by one man sin entered into the world, and death by sin: and so death passed upon all men, for that all have sinned." This teaches that all men die because of Adam's sin and that there was no human death (or sin) before Adam. The fifteenth chapter of I Corinthians clearly teaches that the consequences of Adam's original sin included physical as well as spiritual death. We find in I Corinthians 15:22 that, "For as in Adam all die." Therefore, death came as a consequence of

Adam's original sin. Creationists believe that this refers to both human and animal death. This is because of the fact that the idea of fighting and bloodshed between animals isn't consistent with the concept of a perfectly created world (which God called "very good"). The belief that there was death before the original sin, described in the third chapter of Genesis, seriously undermines the basis of the gospel because it denies the validity of Romans 5:12.

There is some argument over the issue of whether or not the Fall affected the animal kingdom. That is, does the "all die" of I Corinthians 15:22 refer to animals as well as man. Dr. Morris and Dr. Whitcomb discuss this question at length in their classic book, *The Genesis Flood*. Morris and Whitcomb say that the answer is "yes" and give a number of Scripture references to support their position. They offer Romans 8:19-22 as a reference to show that dramatic changes must have occurred in nature as a result of the Edenic curse of Genesis 3. This passage in Romans refers to the entire creation being in a "bondage of corruption" that causes the creation to "groan...and travail...in pain together until now." Genesis 1:30 reads, "To every beast of the earth, and to every bird of the heavens, and to everything that creepeth upon the earth, wherein there is life, I have given every green herb for food." This passage indicates that there were no carnivorous animals in the "pre-Fall" world. Isaiah 11:6-9 illustrates "God's picture of ideal conditions in the animal kingdom." This passage describes various animals, which we would consider to be dangerous predators, as living together peacefully in a world of perfect harmony. Genesis 3:14 indicates that the serpent went through a structural change as a result of God's curse. This was a change far more drastic than that required to change herbivores into carnivores. This change in the serpent's physical structure occurred after the creation week.[11]

According to the Bible, mankind is in need of a Saviour because of the Fall. Therefore, the first three chapters of Genesis are an essential part of the gospel. This is the main reason that this particular portion of Scripture has been under attack for so long. Donald Chittick writes that, "If these chapters are denied or explained away or reinterpreted as myth, the whole foundation of the gospel is removed."[12]

Theistic evolutionists, who attempt to harmonize evolution and Christianity, claim to believe the Bible while also believing that all life has evolved. Theistic evolutionists contradict, and compromise many Biblical essentials.[13] They generally accept the "ape to man" scenario, but they also believe that somewhere along the line God picked out two advanced apes and called them Adam and Eve. Dr. Whitcomb addresses this issue when he writes, "Theistic evolution cannot consistently allow for any physical miracle in Adam's creation. Thus, even after the image of God was put in a male and female ape, their bodies, being unaffected by this spiritual miracle, would continue to be subject to disease and death just like the bodies of other apes. Therefore sin could not be the cause of physical death even in the human race, and Romans 5:12 would be incorrect."[14] In *The Genesis Flood*, Whitcomb and Morris ask, "What are we to say, then, concerning the Fall and the modern science of physical anthropology? We say, on the basis of overwhelming Biblical evidence, that every fossil that man has ever discovered, or ever will be discovered, is a descendant of the super-naturally created Adam and Eve. This is absolutely essential to the entire edifice of Christian theology, and there can simply be no true Christianity without it."[15]

In *Twilight of Evolution* Dr. Morris writes, "Acceptance of evolution is logically followed by the rejection of a high theory of Biblical inspiration, then by rejection of the doctrine of the Fall and the curse, and finally the rejection of the substitutionary atonement". He goes on to say, "The theological capitulation to evolution has been the forerunner and the basis of the development of modernism in religion."[16] So it really does matter what we believe about evolution. If evolution is true, then Genesis cannot be historically and scientifically accurate. If Genesis is not accurate, then we can't really consider it to be inspired by God. Otherwise, God would be guilty of giving mankind false information. If Genesis is not inspired by God, then how do we know that the rest of the Bible is inspired. Ken Ham put it this way when he wrote that, "If the Bible is not the infallible Word of the One who knows everything, then we have exactly nothing. We can never be sure about anything. What then is truth: my word, your word, or someone else's word?"[17]

In II Timothy 3:16 we find that "all Scripture is given by inspiration of God, and is profitable for doctrine, for reproof, for correction, for instruction in righteousness." Genesis must be included in this statement. Jesus made reference to Genesis many times. For example, in Mark 10:7, He quotes Genesis 2:24 where He says, "For this cause shall a man leave his father and mother, and cleave to his wife." In Matthew 24:37, 38, Jesus says, "but as the days of Noe (Noah) were, so shall also the coming of the Son of man be. For as in the days that were before the flood they were eating and drinking, marrying and giving in marriage, until the day that Noe entered into the ark." In Luke 11:51, Jesus refers to Genesis 4:8 when he talks about the "blood of Abel". Another example can be found in John 5:46, 47 where Jesus says "for had ye believed Moses ye would have believed me: for he wrote of me. But if ye believe not his writings, how shall ye believe my words?" In fact, John 5:47 is a pretty good summary of this chapter. Genesis really is foundational to the rest of the Bible.

The rejection of a literal interpretation of the Bible is a consequence of a very destructive domino-effect. According to Talmage, "you follow this crusade against any part of the Bible – first of all, you will give up Genesis, which is as true as Matthew; then you will give up all the historical parts of the Bible; then after awhile, you will give up the miracles; then you will find it convenient to give up the Ten Commandments."[18] Ultimately, the rejection of a literal Genesis leads to the questioning of the most essential matters of doctrine such as the virgin birth of Jesus, the vicarious atonement of Jesus, and resurrection of Jesus. To defend a literal interpretation of Genesis is to defend these same points of doctrine.

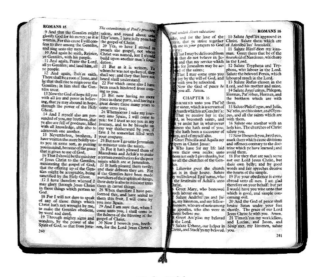

Holy Bible

The eternal, inherent, verbally inspired Word of God

CHAPTER 2

THE BIBLE, SCIENCE, AND FIRST CAUSE

The purpose of this chapter will be to begin to establish the fact that the Bible does have scientific credibility. We will also briefly study Exodus 3:14 which will serve as an introduction to the issue of First Cause. As Henry Morris has written, "one of the most amazing evidences of the divine inspiration of the Bible is its scientific accuracy. There are many unexpected scientific truths that have lain hidden within its pages for thousands of years, only to be recognized and appreciated in recent times."

There are many Bible passages which anticipate scientific discoveries concerning astronomy. For example in Jeremiah 33:22, the Bible states, "the host of heaven cannot be numbered." In Genesis 15:5 we find that Moses wrote, "Look now toward heaven, and tell the stars, if thou be able to number them." Then in Genesis 22:17 we find that the "stars of the heaven" are compared to "the sand which is upon the sea shore." Early astronomers thought that they could actually count the stars. Ptolemy counted 1,056 stars, Hipparchus of Alexandria counted 1,022, and Kepler counted 1,005. Dr. Donald DeYoung, professor of astronomy at Grace College in Indiana, has calculated that there are approximately 10^{22} grains of sand on the Earth. It is startling that this is also the approximate number of stars in all known galaxies. 10^{22} is read as ten billion trillion.[1]

In Job 38:31 we find the question, "Canst thou bind the sweet influences of Pleiades, or loose the bands of Orion?" The term Pleiades is here translated from the word *chimah*, which means pivot or hinge. It has been discovered that the earth, and the rest of our solar system revolves around the brightest star of the Pleiades. This star is called Alcyone and our solar system revolves around this pivot at an estimated speed of 422,000 miles per day. Scholars have known for centuries that Pleiades comes from a word that meant hinge but Job could not have known that it was really a hinge for our entire solar system. Dr. Bill Rice wrote, "Job simply wrote as he was inspired of the Lord."[2]

I Corinthians 15:41 contains another reference to astronomy and says, "One star differeth from another star in glory." To the naked eye, and even through early telescopes, all stars look the same. Except for the sun, they appear as points of light. It has been a fairly recent development for astronomers to recognize that there are many types of stars. In fact, each star has its own unique characteristics.

We find a reference to weather in Ecclesiastes 1:6-7 where the Bible says that "the wind goeth toward the south, and turneth about into the north; it whirleth about continually, and the wind returneth again according to his circuits. All the rivers run into the sea; yet the sea is not full; unto the place from whence the rivers come, thither they return again." This is a strikingly accurate summary of the earth's water cycle.[3] In Job 28:25 we find a reference to "weight for the winds." The book of Job is probably the oldest book of the Bible yet it wasn't until the seventeenth century that men knew that air actually had weight. This is another evidence of the Bible's divine inspiration.

There are a number of scientific truths in the Bible concerning man. In Genesis 2:7 we find that "the Lord God formed man of the dust of the ground, and breathed into his nostrils the breath of life; and man became a living soul. It is only in comparatively recent times that scientists have found that people are composed of the same elements as that of ordinary dust. In Leviticus 17:11, the Scripture says, "The life of the flesh is in the blood." Modern medical science does recognize the importance of blood and goes to great expense to maintain large quantities of blood in storage. Yet blood banks were almost unheard of before World War II. In the early days of our nation, physicians would drain blood out of sick people. These "bloodletters" were common and none other than George Washington died after being bled four times. This doctor was trying to cure Washington's pneumonia.

In Leviticus 13:46, in a reference to contagious disease, we read, "All the days wherein the plague shall be in him he shall be defiled; he shall be unclean: he shall dwell alone; without the camp shall his habitation be." This passage refers to quarantining people with contagious diseases. In Leviticus 13:45 is a reference to the Mosaic requirement for lepers to cover their faces in the presence of uninfected persons. These two verses were written a very long time before the discovery that it is germs that spread disease.

Another example of scientific accuracy in the Bible can be found in Proverbs. In Proverbs 6:6-8 we are instructed to "Go to the ant, thou sluggard; consider her ways, and be wise: which having no guide, overseer, or ruler, provideth her meat in the summer, and gathereth her foot in the harvest." Please notice the word her, which appears three times in this passage. People were probably puzzled about this use of the term her and not his, for centuries. In comparatively recent times, scientists have discovered that when male and female ants embrace, for mating, the male's reproductive organs are torn from his body and he dies. When you see ants out working, you are seeing the females gathering food and making a place to lay her eggs. Dr. Bill Rice wrote, "When Solomon advised us to observe the female ants, he was being scientifically accurate. But how in the world did he know it? Without a microscope, it is absolutely impossible to tell the sex of a little ant!"[4]

As recently as the fifteenth century, most people still believed in a flat earth. Prior to the time of Columbus, even educated people must have considered the Bible to be inaccurate when they read Isaiah 40:22 which refers to the "circle of the earth." They may have felt the same way when they read Proverbs 8:27 which says, "He set a compass upon the face of the deep." This term compass is translated from the Hebrew word *khug* which denotes roundness, or sphericity. Modern scientists (obviously) now agree with the ancient Bible teaching of a round earth.[5]

The Bible even anticipates the second law of thermodynamics in Psalm 102:25-27. In this passage we find that "Of old hast thou laid the foundation of the earth: and the heavens are the work of thy hands. They shall perish, but thou shalt endure: yea, all of them shall wax old like a garment; as a vesture shalt thou change them and they shall be changed; But thou art the same, and thy years shall have no end." This second law is also referred to as the law of increasing entropy. This law basically says that there is a tendency toward deterioration and decay (or aging) at work in the universe. This is what this passage is referring to when it says that the universe "shall wax old like a garment."

To illustrate this principle, let's consider the sun, which is the source of practically all of the earth's energy. However, most of the enormous energy produced by the sun is not directed to the earth and is lost in the sense of being useful energy. In this context entropy is a measure of unavailable energy.[6] Recent discoveries point to the fact that the sun seems to be a young and homogeneous star which has an immense gravitational contraction as its source of energy.[7] It has been calculated that the sun will continue to produce energy for 46 million years by this means.[8] Now 46 million years may sound like a long time but this source of energy is not infinite. The sun, barring supernatural intervention, will eventually burn out. The same principle applies to all stars so that the physical universe is, without question, "running down."[9] We will return to the discussion of entropy later on in this chapter.

Henry Morris has written, "an amazing revelation of modern science is the fact that the physical universe is a tri-universe — a trinity of trinities — perfectly modeling the nature of the Triune God who made it. All true Christians believe in the doctrine of the Trinity: God is one God manifest in three divine Persons — Father, Son, and Holy Spirit." In Romans 1:20 we read, "For the invisible things of him from the creation of the world are clearly seen, being understood by the things that are made, even his eternal power and Godhead." In other words, the creation itself is fashioned according to a triune pattern and we can better understand the Trinity by studying the universe.

The universe itself functions as a space/mass/time continuum. Space can be said to be "the invisible, omnipresent background of all things."

Matter, or mass, is the tangible manifestation of the universe. Mass is what we can actually see and directly experience. Time is another unseen entity through which we live our lives and experience the tangible aspect of the universe. The analogies to the Trinity here are striking.

As in the space/mass/time continuum, only one of the three persons of the Trinity has been tangibly revealed, which is God the Son. The omnipresent background" of space is analogous to God the Father. The invisible entity of time is itself a trinity, consisting of past, present, and future. These three aspects of time are not distinct but form a whole of time. As with the Trinity, only one aspect of time is tangible which is present time. The future is unseen and unexperienced. The past has been experienced but is not now directly tangible. Space is also a trinity in that it consists of three dimensions, which are all equally important and necessary. Each dimension occupies all of space and does not add to space.

When you calculate the volume of a certain area of space you multiply the dimensions instead of adding them. It is an interesting analogy that 1 x 1 x 1 = 1 is also the "mathematics of the divine Trinity." Even matter is a trinity in the sense that it is an extremely concentrated form of unseen energy that manifests itself in motion. This energy can then be experienced as a wide array of phenomena depending on the rates and types of this motion. In matter we have a trinity of energy, motion, and experienced phenomena. So we find that the universe really is a "trinity of trinities."[10]

The Bible makes a somewhat veiled reference to the infinite nature of God in Exodus 3:14 where we read "And God said unto Moses, I Am that I Am: and he said, Thus shalt thou say unto the children of Israel, I Am hath sent me unto you." According to Etienne Gilson, "to say that God 'is this', or that he 'is that', would be to restrict his being to the essences of what 'this' and 'that' are. God 'is', absolutely." He goes on to say, "We do not know what it is for God 'to be'; we only know that it is true to say that God is." So Exodus 3:14 seems to be saying that God has no limitations and is therefore infinite. Gilson also describes the nature of God as "existing itself." Now this statement may sound rather strange but it is basically consistent with Acts 17:28. This verse says that "in him we live, and move, and have our being." To put it another way, we all live, or exist, within another greater Existence which is God.[11]

Let's look at what four Bible commentaries have to say regarding Exodus 3:14. Matthew Henry writes, "This explains his name Jehovah, and signifies that he is self-existent; he has his being of himself, and has no dependence upon any other."[12] When J. F. McLaughlin wrote about Exodus 3:14 he said, "The name (I Am That I Am), whatever its original meaning and use may have been, came to designate for Moses and for Israel the living, self-existent God, the One who is and will be, and who has within himself the exhaustless resources of being."[13] The *KJV Parallel Bible*

Commentary gives the following comments concerning this verse, "God expressed the unchanging, eternal, self-existence of His being. He is able to act at will, to keep promises, to redeem Israel. Yet He is unsearchable."[14] Concerning the two names of God that are given in Exodus 3:14, Spurgeon writes, "By these two names the immutability and self-existence of God are set forth."[15] Notice that all four refer to the self-existence of God. This term means that God does not owe His existence to anyone or anything else. In other words, nothing caused God. This leads us to the issue of First Cause.

When we talk about First Cause, we are talking about the principle of cause and effect traced back in time as far as possible. We know from everyday experience that nothing really happens by itself. All events have a cause, or more accurately causes since each cause must have had a cause. If we try to trace an event back through its causes, we find that we just keep going and going. This study of cause and effect leads eventually to a choice between two alternatives. These two alternatives are: (1) an infinite sequence of secondary causes, or (2) an uncaused primary First Cause which is responsible for creating the universe and ultimately for all things that are in the universe. This First Cause must also be adequate to explain the existence of moral and aesthetic concepts such as justice, righteousness, love, beauty and so on.[16]

Thomas Aquinas deals with the concept of First Cause in his *Summa theologiae*. In that book, Aquinas offers a number of proofs for God's existence. As the second proof, the following argument is presented: "Since nothing can exist prior to and as cause of itself, there must be an order of efficient causes in the world." Aquinas says basically that an infinite series of causes is impossible because such a sequence is without a

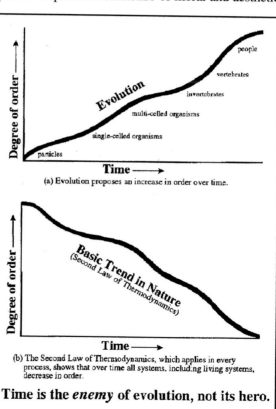

(a) Evolution proposes an increase in order over time.

(b) The Second Law of Thermodynamics, which applies in every process, shows that over time all systems, including living systems, decrease in order.

Time is the *enemy* of evolution, not its hero.

foundation for its very existence. He argued that there must have been a "first efficient cause," which was God and that only God can grant existence to other beings.[17]

At this point it is important to point out that a cause must be sufficient to produce an effect. Dr. Henry Morris addresses this issue in *The Biblical Basis for Modern Science*. In that book he writes that even the "supposed endless chain of finite links can itself be regarded as an effect. Since every component of the chain is a finite effect, the whole series is itself a combined effect, but since the number of links is infinite, its cause must be infinite. Still further, each antecedent link in the chain is 'greater' than the one before it, since something is always lost in the transmission from cause to effect." Dr. Morris is stating here a general principle in the context of creation. This principle is closely related to the law of increasing entropy.[18]

As previously mentioned, all the stars in the universe are in the process of burning out. This inevitable fate of the universe has been called "heat death." This will be a time when all the stars have burned out and all the energy in the universe will have degraded to the same low temperature throughout space. The progression of the universe toward heat death has been called "Time's Arrow." Brown and Weidner address the issue of thermodynamics in their article on physics for *Encyclopaedia Britannica*. They write, "Whereas the total quantity of energy of any isolated system is constant, what might be called the quality of this energy is degraded as the system moves inexorably, through the operation of the law of chance, to states of increasing disorder until it finally reaches the state of maximum disorder (maximum entropy), in which all parts of the system are at the same temperature, and none of the state's energy may be usefully employed. When applied to the universe as a whole, considered as an isolated system, this ultimate chaotic condition has been called the heat death."[20] Dr. Huse wrote, "Given enough time, all the energy of the universe will become random low level heat energy and the universe will have died what is commonly referred to as a heat death." The law of entropy seems to be a result of the curse on the creation which is described in Genesis 3:17-19.[21]

Another aspect of the second law of thermodynamics is that it basically contradicts the Big Bang model of origins. The evolutionists would have us believe that in the very distant past all of the mass in the universe was concentrated in the form of a "cosmic egg." According to Gish, "the size, temperature, and density of this primeval cosmic egg varies according to who is telling the story, but its temperature and density were enormous, while its radius has been estimated to be from no more than that of an electron up to some fraction of a light-year. The cosmic egg was so hot that no elements could exist - the egg consisted of subatomic particles and radiation."[22]

Somehow these mindless subatomic particles are supposed to have begun organizing themselves into higher levels of order at some point after this primeval egg exploded. But an explosion carries no set of instructions and would have resulted in a highly disordered and unstructured universe. Besides that, the principle of increasing entropy says that the universe is going from a high state of order and complexity toward chaos. By contrast, the Big Bang model says that the universe began in a, more or less, chaotic state and is steadily progressing toward higher and higher levels of order. So it is clear that the Big Bang model and the Second Law are completely contradictory.

Another problem with the Big Bang model is that it doesn't explain the "lumps" in the universe. These lumps are the stars and galaxies that are distributed heterogeneously throughout space. A primeval explosion would have resulted in a homogeneous, unorganized distribution of matter. A supposed proof of the Big Bang is that the universe has a uniform, low temperature background microwave radiation.

The evolutionists claim that this radiation, which they say corresponds to about 3 degrees Kelvin, was left over from the Big Bang. However, it has been discovered that this radiation is really quite variable. Even if this radiation was uniform, there are a number of possible explanations.[23] Obviously, no scientist ever observed this supposed Big Bang taking place. Yet a great many people, even highly educated people, fail to realize that the evolutionists are just guessing. They also miss the fact that the Big Bang model is not consistent with the second law of thermodynamics. Gish has written, "The proper understanding of the science of thermodynamics and the theory of evolution from the origin of the universe through the origin of life to the origin of man comes as close as is possible to providing proof that the theory of evolution is scientifically untenable."[24]

The study of entropy has profound theological implications. It is obvious that the universe has not reached heat death. However, without supernatural intervention (see Revelation 21:1-5) it must. Now if Time's Arrow extends infinitely into the past the universe would have already reached heat death. Therefore, Time's Arrow is not infinite at all but must have had a starting point.

Let's look at the first law of thermodynamics which says that "energy can be transferred from one place to another, or transformed from one form to another, but it can neither be created or destroyed."[25] This first law also applies to mass (or matter), because we know from Einstein's equation that matter is a very highly concentrated form of energy; $E=mc^2$. So the total amount of matter plus energy in the universe remains the same. Mass can be converted to energy by nuclear reactions but this is a transformation and not really the creation of anything. The first law basically says that mass cannot create mass and that energy cannot create energy. So we can safely say that the physical universe, being composed of matter and energy, cannot create itself.

Now we are faced with a dilemma. After an examination of the second law of thermodynamics, the only logical conclusion is that there must have been a beginning. But the first law says that the universe cannot create itself. So there must have been a First Cause that was able to create the vast space/mass/time continuum. When we consider the staggering dimensions of space, the enormous yet extremely complex quantities of matter, and the sobering fact of eternity, isn't the infinite, omnipotent, and eternal God of the Bible the only sufficient First Cause of the universe. We must conclude that a wise and benevolent Creator was the First Cause and (as Rice has written) "that there is a God is the basic fact of all this universe."[26]

Planets

CHAPTER 3

DARWIN AND DARWINISM

It is probably fair to say that most people don't have much more than a rather nebulous notion of what evolution really is. This chapter will attempt to show the reader just who Charles Darwin was and also show what his (so-called) theory was all about. Before we look at Darwin's life, let's begin with four definitions of evolution.

The late John R. Rice defined evolution as the "natural development of animal and vegetable life from dead matter; and of the many-celled form coming from a one-celled form; and of the coming of eyes and ears and feet and hands and liver and lungs and digestive apparatus and heart into being by natural processes. And so the evolutionary theory is that reptiles come from fish, birds from reptiles, mammals from birds, and man from mammal."[1] Dr. Gish defines evolution as, "the theory that all living things have arisen by a materialistic evolutionary process from a single source which itself arose by a similar process from a dead, inanimate world. This theory may also be called the molecule-to-man theory of evolution."[2]

Phillip Johnson has written that, "evolution in Darwinist usage implies a completely naturalistic metaphysical system, in which matter evolved to its present state of organized complexity without any participation by a Creator."[3] Finally, Scott Huse calls evolution, "an imaginary process by which nature is said to continually improve itself through gradual development."[4]

The name most closely associated with the term evolution is, of course, that of Charles Robert Darwin. He was born in Shrewsbury, England on February 12, 1809. As the son of a wealthy physician, he had the advantages of an

Charles Robert Darwin

aristocratic background. As a youth, Darwin attended Shrewsbury School for seven years, and at age 16 he began to study medicine at the University of Edinburgh. While in Scotland, Darwin met a man named Robert Jameson who kindled an interest in historical geology. Darwin was a poor medical student, especially disliking the operations.

After dropping out of Edinburgh, he was sent by his father to study theology at Cambridge in 1827. Darwin, at this point in his life, still held the beliefs of the Church of England. Although he was not an outstanding student, he did earn the BA degree in 1831. While at Cambridge, Darwin became friends with many scientists who encouraged his interest in science. A few months after leaving Cambridge, he became an unpaid naturalist on the HMS *Beagle*. Darwin was invited upon the recommendation of botanist John Stevens Henslow, whom he had known at Cambridge.

The voyage was for the purpose of surveying the coasts of South America, and to establish a chain of chronometric stations in the Pacific Islands. The *Beagle* left Plymouth, England on December 27, 1831 for a five year voyage during which Darwin sent back biological and geological specimens. He wrote three books on South American geology from notes he took during this voyage.

After returning to England in 1836, Darwin began a series of notebooks on what he called the species problem. In these, he collected facts through correspondence and reading, as well as discussions with gardeners, naturalists, and breeders. These notebooks were for the purpose of gathering evidence to support his ideas on evolution.

An essay on population by Thomas Malthus had a great impact on Darwin's developing ideas on organic evolution. Malthus essentially said that population is held in check by a limiting food supply. Darwin was particularly interested in the competition between individuals of the same species. He saw that different characteristics enhanced an individual's survival depending on the environment. Darwin claimed that these advantageous traits were passed to succeeding generations by a process known as natural selection.[5]

Darwin's process of natural selection is also referred to as Darwinism, which has the idea of survival of the fittest as its basis. He observed the natural world's near constant struggle for food. He reasoned that those individuals possessing characteristics that would give them an advantage in this struggle for survival would be more likely to reproduce. For example, Darwin saw the long neck of the giraffe as a type of survival advantage that had been developed by natural selection. However, as Scott Huse writes, "natural selection explains survival of the fittest, but it does not explain arrival of the fittest."[6]

The fact that natural selection occurs is not really in question. Phillip Johnson writes that, "species avoid genetic deterioration due to natural attrition among the genetically unfit." Individuals, of any species, having serious birth defects usually do not survive to maturity, and therefore don't reproduce. But the claims of Darwinism go far beyond this. Charles Darwin supposed that natural selection would keep improving a species until it became an entirely different species. He believed that a unicellular organism could become a highly complex organism by this process of natural selection if given enough time (i.e. millions of years). Darwinists say that these improvements are due to beneficial mutations, which are genetic changes that occur randomly. They claim that a small fraction of these mutations are beneficial to the organism and improve its survivability. Individuals receiving this genetic improvement will therefore be more likely to reproduce, preserving this beneficial mutation. This new trait will supposedly spread throughout and become a characteristic of an entire species. According to evolutionist dogma, if there are enough beneficial mutations of the right kind, then over time, new organs and patterns of behavior will result.[7]

Darwin wrote that, "it may metaphorically be said that natural selection is daily and hourly scrutinizing throughout the world, the slightest variations; rejecting those that are bad, preserving and adding up all that are good; silently and insensibly working, whenever and wherever opportunity offers, at the improvement of each organic being in relation to its organic and inorganic conditions of life. We see nothing of these slow changes in process, until the hand of time has marked the lapse of the ages."[8]

Dr. Chittick addresses the mutation issue when he writes that, "one organism which has been studied very extensively is the fruit fly. Many generations of fruit flies have been raised in the laboratory. They have been artificially mutated time after time. And yet fruit flies refuse to become anything but fruit flies. Observation in the laboratory simply does not support Darwinism or neo-Darwinism."[9]

In his writings, Darwin would begin by discussing the importance of selection in domestic plants and animals. Then he would compare domestic selection to natural selection.[10] However, natural selection can't really be equated with artificial selection because they are basically two different things. Domestic breeding produces characteristics for purposes other than survival. There is no question that when domesticated animals are sent into the wild, that the very specialized breeds do not survive. The breeds that do survive soon revert to the very same wild type that human breeders began to domesticate in the first place. Natural selection, in this context, actually becomes a "conservative force" that really limits variation.

The observation of artificial breeding shows that many, many generations of careful selection produces no new species. It also shows that there are limits to the variation that breeding can produce. A particular species has fixed genetic limits for change. Take the breeding of dogs, for example. After thousands of years of breeding, you don't see dogs as big as elephants. Dogs, as other animals, can be bred for size until a certain upper genetic limit is reached, No amount of breeding can take them beyond that limit. This limit is established in what can be called the gene pool. However, Darwinists would argue that random mutations will alter the gene pool and therefore increase this capacity, but they cannot prove that such a thing has ever really occurred.[11]

It must be understood that Charles Darwin was virtually ignorant of the principles of genetics when he wrote *The Origin of Species*. In fact, Gregor Mendel's pioneering work on genetics was largely ignored when it was published in 1866. Darwin was completely mistaken in his ideas on the passing of acquired characteristics from generation to generation. He thought that hereditary factors called gemmules, are formed in tissue cells when these cells are affected by the environment. It was thought that these gemmules were somehow transported to germ cells, and then inherited by offspring. However, modern scientists know that there is no such thing as a gemmule, and that there is no inheritance of acquired characteristics.

Inheritance is actually controlled by elements called genes which are only found in the germ cells. Germ cells include the ova (or eggs) and sperm. A gene actually consists of a long strand of subunits, called nucleotides. Deoxyribonucleic acid (DNA) is the very complicated chemical of which genes are comprised. There are four types of nucleotides and it is their order that identifies a particular gene. An inherited trait is determined by at least two genes, one coming from each parent. This pair of genes is called an allele. At fertilization, the single sets of genes contained in the egg and sperm are combined. An individual's sperm, or egg, cells are not equivalent. That is, they have different gene combinations which result in great genetic diversity, depending on which sperm fertilizes which egg.

A given gene normally exists for a very long time without any structural change. But on rare occasion, exposure to an outside agent (e.g. x-rays, ultraviolet light, cosmic rays, and various chemicals) can produce a structural change in a gene. These changes are referred to as mutations. Those mutations that have either been induced under laboratory conditions, or observed in nature have always been harmful. But evolutionists still claim that about 0.01% of mutations are beneficial to the plant or animal. This claim is made because evolutionist dogma requires beneficial mutations and not because of observed scientific data.[12]

Another unscientific notion that is essential to Darwinism is that of spontaneous generation. This concept was a common belief during the middle ages.[13] During those times ignorant people thought that spoiled meat would spontaneously produce maggots and that germs spontaneously arose in water. Those people thought that virtually all life arose in this manner. Then Pasteur proved that life never arises from non-life. By placing boiled water in sealed containers, he demonstrated that this water would remain germ-free indefinitely.[14] In another experiment which involved the filtration of air, Pasteur proved that putrefaction of food is caused by germs present in the air and that such food doesn't spontaneously generate new organisms within itself, as a result of putrefaction.[15]

A third unscientific concept that is essential to evolutionary dogma is the idea that there must be a natural tendency from disorder to order. According to the evolutionists, this is what has happened to the universe in its journey from "primordial chaos to the cosmos." They contend that galaxies, stars, solar systems, plants, animals, and even humans basically created themselves due to this supposed tendency toward self-organization. But has such a tendency actually been observed?

The answer is no. Dr. Gish writes, "No scientist has ever detected a tendency of matter towards self-organization. Matter does not tend to promote itself from disorder to order." The tendency of matter actually goes in the opposite direction and is referred to as the Second Law of Thermodynamics. This tendency is also referred to as the law of increasing entropy.[16] Entropy is basically a measure of randomness or disorganization. A state of low entropy is associated with orderliness and a state of high entropy is one of randomness.[17] Entropy can also refer to the useful energy in a system. As entropy increases, the amount of energy available for work decreases even though total energy remains the same. The universe is not only moving toward a higher state of disorder but is also running down (so to speak) in terms of available energy.[18] So we find that the concept of evolution contradicts the Second Law of Thermodynamics.[19]

The observed evidence really points to a period during which a high degree of organization and energization were introduced into the Universe. This evidence is entirely consistent with the creation model of origins. The Bible describes such a time in Genesis, Chapter 1. In Genesis 2:1-3, we read that this unique period of history was brought to a definite end. Some evolutionists have attempted to harmonize entropy with evolution. However, Whitcomb and Morris write that this is "an attempt to reconcile it (the law of entropy) with that with which it is utterly irreconcilable, the assumption of universal developmental evolution!". They also point out, "The one is itself the negation of the other." The law, or laws, of creation that operated during the creation week have been replaced by this law of increasing entropy which says that things just sort of naturally deteriorate. The law of entropy probably came to be as a part of the Edenic curse as described in Genesis Chapter 3 (but back to Darwin).[20]

Charles Darwin was strongly influenced by a man named Charles Lyell, who was a Scottish lawyer. Lyell was a proponent of the principle of uniformitarianism. This principle basically explains all geological formations as the products of very slow processes which act over long ages of geological time. Uniformitarianism is foundational to evolutionary biology. Lyell claimed that because these physical, chemical, and biological processes act so slowly, the Earth's physical features must have developed over an enormous span of time.[21] Dr. Scott Huse defines uniformitarianism as, "the concept that the present is the key to the past. Processes now operating to modify the earth's surface are believed to have been operating similarly in the geologic past; that there is a uniformity of processes past and present."[22]

Although he was an Oxford educated lawyer, Lyell was a geologist by avocation. He published his well known *Principles of Geology* in 1830, and to a great extent, spent the rest of his life revising and updating that book. Darwin took the first volume of Lyell's book on the Beagle. By 1835, he was apparently a complete convert to the concept of uniformitarianism. By contrast, Darwin's earliest geological training had been by Jameson who was a catastrophist. The influence of this training could be seen early in the voyage when Darwin wrote that he saw evidence of a great flood in South American geology.[23]

Catastrophism has been defined as "the belief that the fossils, rock formations and other features of the earth's crust have been formed rapidly in a relatively short period of time during a worldwide disaster."[24] In fact, that disaster was widely believed to have been the Genesis flood up until the time of Darwin and Lyell. This had been true since about the time of Galileo and Kepler. Dr. Morris writes in *The Genesis Flood*, "Before 1800, some of the outstanding theologians of the Church were of the opinion that the Genesis flood was not only universal in extent but also was responsible for the reshaping of the earth's surface, including the formation of sedimentary strata. Among those who held this view were Tertullian, Augustine, Chrysostom, and Luther." Uniformitarianism was clearly an attack on the concept of a universal flood as being responsible for shaping the Earth's surface.[26] John R. Rice wrote, "Benjamin Silliman, head of the Geology Department of Yale University in 1829; and many other scientists - Williams, Cattcot, Woodward, Granville Penn - held what we call flood geology, believing that the Genesis flood and the adjustment period in the first century or two afterward are accountable for the geologic conditions of the surface of the earth today."[27]

Some believe that the doctrine of uniformitarianism was actually a fulfillment of New Testament prophecy. II Peter 3:3-6 reads as follows, "Knowing this first, that there shall come in the last days scoffers, walking after their own lusts, And saying, where is the promise of his coming? for since the fathers fell asleep, all things continue as they were from the beginning of the creation. For this they willingly are ignorant of, that by the word of God the heavens were of old, and the earth standing out of the water and in the water: Whereby the world that then was, being overflowed with water perished." This passage indicates that evolutionist scientists disbelieve the Bible as a matter of personal choice. These evolutionists don't want to believe in a cataclysmic Genesis flood so they structure their theories to explain it away. The part of this passage that says, "all things continue as they were from the beginning" is a pretty good definition of uniformitarianism. This was exactly the idea that Lyell was trying to make people believe.[28]

It needs to be mentioned that evolution is actually a very ancient concept that did not originate with Charles Robert Darwin. For example, there were evolutionary beliefs among the early Greeks. Empedocles believed that animals had somehow evolved from plants and a fish-to-man evolution scenario was taught by Amaximander. Henry Morris has written that, "it was commonly thought that not only insects and fishes, but probably also the higher animals and man himself, were on occasions generated directly from mud or slime or some other inorganic medium. And if such great marvels as this could and did occur, there was no great problem in believing that one species could be transmuted into another."

All of the ancient creation myths from the non-Christian religions follow a type of evolutionary scenario. That is, there is some type of primeval system, or primeval chaos, upon which the "gods" act to produce the present world and its inhabitants. Therefore, ancient peoples were familiar with evolutionary concepts concerning creation. It has been said that Genesis does not present the true picture of creation (i.e. evolution) because the early Hebrews would not have been able to comprehend it. But this is a rather ridiculous position to take since early peoples were already thinking in evolutionary terms.[29]

But where did the modern version of evolution come from? Dr. Morris writes in *The Long War Against God*, "Evolutionism was in significant part carried into the modern world through the ancient tradition of the Great Chain of Being."[30] This Chain of Being is said to have been the "seed of evolution", and was basically equivalent to the *Scala Naturae* of Aristotle. It was a fashionable concept among educated people from the Renaissance until about the end of the eighteenth century. According to this "quasitheological theory" a linear scale can be constructed to show increasing complexity among living organisms. Man was placed at the top with the very simple organisms on the bottom. This scale also included earth, air, water, metals and stones beneath the simplest organisms. This chain extended above man where God and the other celestial beings were represented.[31]

As Lovejoy has written, "the graded series of creatures down which the divine life in its overflow had descended might be conceived to constitute also the stages of man's ascent to the divine life," (i.e. an evolutionary type scale) Lovejoy believed that Thomas Aquinas was referring to this chain when he wrote, "The perfection of the universe therefore requires not only a multitude of individuals, but also diverse kinds, and therefore diverse grades of things." Notice the phrase "grades of things".[32] Thomas Aquinas has been credited with introducing the Chain of Being concept, along with other aspects of Aristotelian philosophy into the Catholic church.[33]

There are numerous references to this concept in literature. For example, Milton referred to the chain in *Paradise Lost* where he wrote

The scale of nature set
From center to circumference, whereon
In contemplation of created things
By steps we may ascend to God.[34]

James Thomson refers to this chain in *The Seasons* where he wrote

Has any seen
The mighty chain of being, lessening down
From Infinite Perfection to the brink
Of dreary nothing, desolate abyss!
From which astonished thought, recoiling, turns?

Alexander Pope made the following reference to this chain, or graded scale, in his *Essay on Man* (epistle one)

Vast chain of being! which from God began
Natures aethereal human, angel, man
Beast, bird, fish, insect, what no eye can see
No glass can reach, from infinite to thee,[35]

As a final example, Edward Young praises the continuity of the chain, seeing it as a proof of the soul's immortality, in *Night Thoughts* where he writes

Look Nature through, 'tis neat gradation all.
By what minute degrees her scale extends!
Each middle nature join'd at each extreme
To that above it, jointed to that beneath
But how preserved
The chain unbroken upwards, to the realms
of Incorporeal life? those realms of bliss
Where death hath no dominion? Grant a make
Half-mortal, half-immortal; earthy part,
And part ethereal; grant the soul of Man
Eternal, or in Man the series ends.[36]

The supposed fossil sequences of the geologic column, as it appears in science textbooks, were really an adaptation of the Chain of Being. The geologic column was based, in reality, on this chain concept and not on any observed succession of fossils in the layers of rock.[37] Concerning this same topic, Morris has written, "It would be a more or less natural exercise for philosophers whether ancient or modern - to try to organize living things in order of increasing complexity from microbes to men. This arbitrary sequence, with all its very real gaps, constitutes the only factual basis for the Great Chain of Being and for the evolutionary ladder that was based on it."[38] As Lubenow has written, "Like the keys on a piano, each organism was discrete but a bit higher in the organization and more complex than the one below it." The chain concept bore so much resemblance to an evolutionary progression that it was easily converted. The Great Chain of Being was so well known that it served as a preparation for Darwinism.[39]

The first modern evolutionists appeared in France in the eighteenth century. Their names were Pierre de Maupertuis, Benoit de Maillet, and Comte de Buffon. The mathematician Maupertuis was like his friend Voltaire in that he was an enemy of Christianity. The evolutionary ideas that he advocated were anti-theistic by design. In his writings, Maupertuis addressed the issues of mutations, genetic heredity, and natural selection. De Maillet believed in organic evolution and that the earth was infinite in age. He believed in a uniformitarian development and also wrote about the occult. Comte de Buffon was, for many years, in charge of the French Royal Botanical Gardens. In his massive work, *Histoire Naturelle*, he anticipated most of the later evolutionary ideas of Charles Darwin.[40]

Jean-Baptiste de Lamarck was another French evolutionist who came along after the other three. He is said to have produced the first "systematized theory" of evolution. He believed that the bodily organs of any species would become developed, more or less, according to use. The idea that the giraffe gained an elongated neck due to constantly reaching for food and then passing this acquired characteristic on to offspring is an example of Lamarckism.[41] A second example is that the children of a blacksmith will have stronger arms than other children. According to Lamarckism, characteristics that are acquired by parents can be passed on to their offspring.[42]

Lamarckism was scientifically discredited as scientists began to understand that characteristics, passed on to offspring were determined by the DNA contained within the genes of parents. Charles Darwin had very similar ideas to Lamarck concerning traits acquired by use. Darwin supposed that there had been a drought on the African prairies long ago which initiated a great competition for food. As this drought continued giraffes were always stretching for food in higher and higher branches and consequently gained an elongated neck. Darwin thought that each organ produced "pangenes" which flowed through the circulatory system to the reproductive organs. He also supposed that bigger organs would produce more pangenes (e.g. the big arms of the blacksmith) and that these pangenes were transferred via the male's semen to the female.[43]

Another early proponent of evolution was Charles Darwin's own grandfather, Erasmus Darwin. Erasmus was a physician and also a naturalist. He was, roughly, a contemporary of Lamarck. The term Darwinism was actually first applied to the evolutionary ideas of Erasmus, as they appeared in his book *Zoonomia*. This book, which was published in 1794, may have been an inspiration for Lamarckism. *Zoonomia* offered a theory of natural selection which was probably largely plagiarized by Charles in *The Origin of Species*.[44] However, in Charles' defense, this may have been due to the fact that Erasmus was friendly with many men who were known sympathizers to the French Revolution. It has been said, "The reaction in England, to the French Revolution was destined to sweep Erasmus Darwin's ideas out of fashion." Therefore, Charles probably did not want the ideas expressed in *The Origin of Species* to be associated with his controversial grandfather.[45]

Two creationists probably also influenced Charles Darwin, as far as his views on natural selection were concerned. One was a theologian named William Paley. Darwin had become familiar with Paley's writings while studying at Cambridge. Paley wrote about natural selection as a means to remove unfit individuals and prevent them from passing on inferior characteristics to subsequent generations.[46] Another creationist who seems to have influenced Darwin was a man named Edward Blyth. As a matter of fact, Blyth published his concept of natural selection 24 years before *The Origin of Species* was published. Blyth saw natural selection as a means by which a created species could adapt to an environmental change.[47] Blyth was a London chemist who often gave lectures at meetings of London scientists. As Hitching has written, "Darwin's early notebooks on 'transmutation' (for years he avoided using the word evolution) contain transcriptions of what Blyth said. In 1835 and 1837 Blyth's theories were published in the British Magazine of Natural History. Darwin, according to a cryptic reference in a letter, seems to have read these too."[48]

Alfred Russel Wallace was an evolutionist who was a contemporary of Charles Darwin. Darwin and Wallace have been credited as being "co-discoverers" of natural selection, even though neither of them really were. Wallace was a naturalist who spent much of his life exploring Malaysia and the Amazon. Wallace was in Malaysia in 1858 when he became very ill with malaria. It was during this illness that Wallace wrote a paper that was virtually identical to Darwin's own ideas on natural selection. Wallace had already corresponded with Darwin and he sent this paper to him, asking

Darwin to review it. Wallace was hoping that he might recommend it for publication.

After reading Wallace's paper, Darwin immediately contacted Charles Lyell. Lyell strongly urged Darwin to get busy and finish writing the book about evolution,[49] which he had begun in May, 1856. Lyell also arranged that Wallace's paper, entitled *On the Tendency of Varieties to Depart Indefinitely from the Original Type* to be read at the next meeting of the Linnaean Society. An outline of natural selection written by Darwin in 1844 was also read at that meeting which ostensibly gave Darwin priority to the theory.[50] This meeting of the Linnaen Society was in July of 1858. With the guidance and encouragement of Lyell, Darwin began condensing what had been done on his original book which had been tentatively called *Natural Selection*. *The Origin of Species* was of a length that would appeal to a wider audience and was published in November of 1859.

The influence of Lyell, on Charles Darwin, should not be underestimated. He was really almost a father figure to Darwin. *The Origin of Species* might not have been published without Lyell's influence. His principle of uniformitarianisn provided the long ages of geologic time that Darwinism required. Uniformitarianism and Darwinism cannot really be separated.[53] They work together to undermine the credibility of Genesis. Darwinism is an attack on the Biblical account of creation. Uniformitarianism is an attempt to explain the geologic features of the Earth as being the result of extremely slow processes instead of the global diluvial catastrophe that is described in Genesis 7 and 8.[51]

Christians are warned, in I Timothy 6:20, about "oppositions of science falsely so called." Christians are also warned about "cunningly devised fables" in II Peter 1:16. The concept of evolution, including its most popular version called Darwinism, is both false science and fable. Darwinism is false science because it rests upon a pseudo-scientific basis of spontaneous generation, beneficial mutations which have not been observed, missing links which have not been found, and an annulment of the second law of thermodynamics. Evolution is a fable in that it is drawn mostly from speculation and conjecture. It is also a fable because it has a basis in ancient mythology. Dr. Morris wrote, "Modern 'scientific' evolutionism is not new at all, but merely an updated, and somewhat more sophisticated version of ancient cosmogonic myths."[52]

Lord Kelvin

George Washington Carver

Wernher von Braun

CHAPTER 4

CHRISTIAN MEN OF SCIENCE

Many people have the notion that the Bible is in conflict with science and that (as a consequence of this conflict) real scientists just don't believe the Bible. There is actually no such conflict. Further more, modern science had its origin in a predominantly Christian culture,[1] and had the Bible as its philosophical foundation.

Dr. Chittick refers to this foundation in his book, *The Controversy*, where he writes, "A created universe was expected to have design, order, and purpose. Man, using his rational mind, could study this ordered universe in a rational way and seek to discover its laws; and modern science is based on the assumption of scientific law." In other words, the Bible gives us a kind of framework in which the universe was created by a "reasonable God" whose scientific laws can be investigated and understood by reasoning men. Chittick goes on to say that this "explains why science did not develop in the Eastern countries with their materialist philosophies and pantheistic religions. They simply did not have the proper philosophical base." If the universe did come about as the result of chance interactions of matter and energy, then such a universe would not be expected to function in a rational way. The mind of man would also have been produced by chance interactions and therefore would not be capable of deliberate, logical, and rational scientific study.[2]

Morris has written in his book, *The Long War Against God*, "The 'scientific revolution' did not take place until the way for it was prepared by the Reformation and the Great Awakening in western Europe and North America, with the great upsurge in Bible study and evangelical Christianity that followed." He goes on to say in that same book that "some of the most incisive modern thinkers have recognized that science could only have arisen in a creationist context."[3] Chittick states categorically, "Creation is the foundation on which modern science began."[4]

A Biblical world view is necessary for empirical science. This world view is consistent with the law of causality. According to this law of causality, like causes have like results and particular, predictable effects follow from particular non-variable causes. The Biblical world view includes the concept that we can expect natural phenomena to follow established laws which can be investigated and defined by means of man's rational mind. Science is really based upon the premise of a sentient and rational

First Cause who created and now controls the Universe by means of scientific laws which He enacted and now enforces (See Hebrews 1:3). The idea that a real scientist cannot be a creationist is an absurdity. In fact, the majority of the great pioneers of science believed in the great tenets of the Bible, including Biblical creation.[5] Following is a series of short biographies of many of these Christian scientists.

Leonardo da Vinci (1452-1519), who was a pioneer in the field of physics, especially hydraulics,[6] is even said to be "the real founder of modern science." Besides being a great artist, he was also an architect and engineer. da Vinci was the designer of a number of public works projects in Milan. He compiled extensive notebooks in fields such as biology, anatomy, physics, optics, and aeronautics. da Vinci was an advocate of empirical research long before the development of the scientific method. da Vinci was a gracious and sincere Christian who lived according to a high standard of morality.[7]

Johann Kepler (1571-1630) is credited with establishing the field of physical astronomy. He was able to show that the earth and other planets actually revolved around the sun. Kepler also contributed to the development of calculus. He was a dedicated Christian who had two years of seminary study before entering the field of astronomy.[8] His study of planetary motion led him to the discovery that planets move in elliptical orbits rather than circular. In 1594, he was appointed to be a mathematics teacher in Graz, Austria with the additional duties of preparing almanacs for the Province of Styria. It was about this time that he began the serious study of astronomy and the result was his first book, *The Mysterium Cosmographicum*. This book, published in 1596, was an attempt to explain the principles by which the universe was created by God.[9]

Francis Bacon (1561-1626), who held the title of Lord Chancellor of England, was a contemporary of Kepler. Bacon is credited with establishing and formulating the scientific method. He was a Bible believing Christian who recognized that induction, based on experimentation, was necessary in scientific investigation.[10] Bacon sought to take science out of the realm of Aristotelian deduction and place it on the sure foundation of empirical research. The term empiricism refers to the search for knowledge by observation and experiment. Bacon was perhaps the first to advocate the use of carefully controlled environments for experiments.[11] As opposed to Aristotle's deduction from accepted axioms, the scientific method of Bacon involved the process of induction. Bacon

Francis Bacon

said that scientists must begin to formulate general laws based on observed experimental data and then attempt to find supporting evidence from further experimentation.[12]

Blaise Pascal (1623-1662) was a creationist who established the scientific field of hydrostatics and helped to found hydrodynamics. He is acknowledged as being one of the greatest mathematicians of his era for laying the foundation of differential calculus and the theory of probability among many other contributions. Pascal was an early advocate of the scientific method. He was a devout Christian who wrote a religious work called *Pensees*.[13] These writings, published posthumously, were originally called *Apology for the Christian Religion*. An apology, in this context, refers to a defense of the Christian faith. It has been said that Pascal's exegesis of Scripture in *Pensees* was fundamentalist.[14]

Robert Boyle (1627-1691) is generally considered to be the "father of modern chemistry." He discovered the basic principles of gas dynamics, which is the relationship between temperature, volume, and gas pressure. He made numerous important contributions in physics and chemistry. Boyle was one of the founders of the Royal Society of scientists. He was a strong defender of the Christian faith and was known to have donated much of his personal wealth to the work of Bible translation.[15]

John Ray (1627-1705) was a Christian scientist who has been credited with being the "father of English natural history". He was an important taxonomist, especially as far as English flora was concerned. Ray was a recognized authority on both botany and zoology. He was a creationist who wrote on natural theology. Natural theology refers basically to arguments for some religious truth, such as the existence of God, by a rational study of the natural world. His principal work was *The Wisdom of God Manifested in the Works of the Creation*. He also opposed deistic evolutionists, such as Descartes.[16]

Nicolaus Steno (1631-1686), who was born in Denmark, made very important contributions to the field of stratigraphy. Morris has written, "Steno, with his extensive field studies, developed the principles of stratigraphical interpretation which are still considered basic today." He interpreted the various strata, with their multitudes of fossils, as being the result of the Genesis Flood. To Steno, the fossils represented the remains of living creatures and plant life which died during that great global catastrophe. He took up a religious vocation later in life and wrote extensively on theology.[17]

John Woodward (1665-1728) was a Christian who was one of the founders of geology as a scientific discipline. Woodward established a great museum of paleontology at Cambridge and also did research on soil fertility. He held the Bible in very high esteem and wrote a book titled *Essay*

Towards a Natural History of the Earth which was both Biblically and scientifically sound. In this work, Woodward advocated Flood geology,[18] and his book was very influential in western Europe.

Thomas Burnet (1635-1715) was an English geologist and clergyman who is remembered for writing *Sacred Theory of the Earth*. This was another book advocating the idea that the strata and fossils were produced by the flood of Genesis, chapters 7 and 8 (i.e. Flood geology). In this work, Burnet presented the Bible as a kind of framework for interpreting earth history. Burnet, along with Woodward and William Whiston promoted the "Flood theory of geology" in the latter part of the seventeenth century.[19]

William Whiston (1667-1752) was a Cambridge University scholar who wrote a book called *A New Theory of the Earth*. This book was an attempt to interpret the Genesis record of creation with new findings in geology and physics. This book may have had a weakness in that Whiston leaned toward a naturalistic explanation for the various phenomena involved. However, he did acknowledge that the Biblical record of creation was accurate as far as a recent creation and the Flood were concerned. Whiston dedicated this book to Sir Isaac Newton. It is remarkable that Newton was also an advocate of a recent creation. Isaac Newton actually wrote a book in which he defended the Ussher Chronology, which places the time of creation in the year 4004 B.C.[20]

Carolus Linnaeus (1707-1778) was a Swedish taxonomist whose classification system is still in use today. Linnaeus is said to have been a "man of great piety and respect for the Scriptures." The motive for his comprehensive classification of living things was to construct an outline of the kinds of Genesis 1:11, 12, 24, and 25. The Linnaean species category was meant to be the equivalent of the Biblical kind. However, he made some mistakes. He believed that there was variation within a kind but didn't accept the idea that one kind could evolve into another kind. Linnaeus believed that variation occurred within fixed limits.[21]

Increase Mather (1639-1723) was a clergyman in colonial New England. He earned a bachelor's degree from Harvard at age 17 and also studied in Ireland. Although he was not a scientist, strictly speaking, he has been called an "avocational astronomer and promoter of science." His main astronomical interest seems to have been comets which he avidly studied and wrote about. He founded the Philosophical Society and also served as the president of Harvard College from 1684 to 1701.[22]

Gottfried Wilhelm Leibnitz (1646-1716) was a contemporary of Isaac Newton. He is credited with being the co-discoverer of calculus, along with Newton. Besides many scientific and mathematical accomplishments, Leibnitz was a deeply religious man who wrote about and defended the doctrine of the Trinity.[23] He combined theology and philosophy in a work called *Theodicy*, where he addresses the apparent contradiction between a good God and the presence of evil in the world. Leibnitz believed in the concept of a good God who does not will evil but does permit its existence.[24]

John Flamsteed (1646-1719) was a clergyman who was England's first Astronomer Royal. This Christian astronomer founded the renowned Greenwich observatory. Flamsteed is credited with producing the first extensive map of the stars done by telescope. As a result of Flamsteed's observation, the Earth's meridians are referenced to Greenwich. His observatory was 0° longitude. He observed and catalogued stars from Greenwich for almost three decades.[25]

Jonathan Edwards (1703-1757) was a pastor, missionary, and educator who also showed a remarkable aptitude for science at a young age. Morris has written, "Edwards exhibited deep understanding and original insights into physics, meteorology, and astronomy, far in advance of his time". Edwards probably had the ability to become an outstanding scientist but God took his life in another direction.[26] He graduated from Yale in 1720 as class valedictorian and later earned a graduate degree in theology at that institution. He pastored a large church in Northampton, Massachusetts and was among the leadership of the Great Awakening. Edwards is perhaps best remembered for his famous sermon; *Sinners In the Hands of an Angry God.*[27]

Sir William Herschel (1738-1822) was another great Christian astronomer. He made the finest reflecting telescopes of that era, some of which were quite large. Herschel is probably best remembered for discovering Uranus and for being the first to find double stars. He did the most extensive telescopic survey of the stars up to that time. Herschel was a member of the Royal Society and deeply believed that the universe was a testimony to God's creative power. He solved the nebulae problem with his powerful telescopes which allowed him to see that they were really great collections of individual stars.

Timothy Dwight (1752-1817) was an educator who is remembered for his influential ministry of Christian apologetics. He wrote about Flood geology and could persuasively relate science to Scripture. Dwight was the president of Yale College from 1795-1817. During his years at Yale, he preached an important series of chapel sermons in which he presented "Scientific Christian Apologetics" to the student body and faculty. Due to Dwight's profound Christian influence, large numbers of Yale students trusted Christ for salvation.

Benjamin Silliman (1779-1864) was an outstanding geologist who was also a creationist. Silliman attended Yale and may have trusted Christ as

Saviour due to Timothy Dwight's influence. He later joined the faculty at Yale and founded the Sheffield Scientific School there. He also made contributions to the field of mineralogy and was the first president of the Association of American Geologists. He founded the *American Journal of Science*, which is still an important geological publication.[29]

Humphrey Davy (1778-1829) was an English Christian who was recognized as one of the great chemists of his time. Davy is credited with being an inspiration for Michael Faraday to pursue a career as a scientist. Sir Humphrey isolated a number of chemical elements for the first time such as sodium and potassium. His most notable contribution was probably the development of the miner's safety lamp, and he also demonstrated that the diamond is a highly condensed form of carbon.

Michael Faraday (1791-1867) was a fundamentalist Christian who was born in England. He was a scientific pioneer in the field of physics. Faraday is credited with the discovery of electromagnetic induction, the invention of the generator and many other important scientific contributions in physics and chemistry. He was a man of prayer with strong confidence in the Bible. Faraday took for granted the fact that the Bible and genuine science were in agreement. He served as a laboratory assistant for Sir Humphrey Davy early in his career.[30]

Jedidiah Morse (1761-1826) was a Congregational minister who was also a leading geographer. He wrote the first American geography text which was widely used for decades. Jedidiah was the father of Samuel F.B. Morse. The elder Morse was a creationist who strongly advocated Flood geology. His textbook had a section which addressed the issue of the preservation of animal life on Noah's ark and the subsequent distribution of these animals following the Flood. This is an issue which will be discussed in chapter five.

Benjamin Barton (1766-1815) was a prominent Christian physician who was also an authority on botany and zoology. He was on the faculty at the University of Pennsylvania and wrote America's first botany textbook. He believed that there had been a relatively recent creation in accord with Scripture. Barton was also a student of Biblical ethnology, especially the origin and distribution of the various tribes and nations. He defended the concept of the "unity of the human race" as taught in the Bible and that the nations were originally dispersed from Ararat.[31]

Samuel F.B. Morse (1791-1872) is probably most famous for inventing the telegraph. The very first telegraph message, which was, "What hath God wrought!", can be found in Numbers 23:23. This reflected Morse's desire to give God honor in all matters. Morse graduated from Yale in 1810 and was greatly influenced by Timothy Dwight while he was there.

Another important contribution made by Morse was that he built the first camera in this country. He was also an accomplished artist and studied in London for a few years. He later served as Professor of Sculpture and Painting at New York University. Samuel F.B. Morse firmly believed that the Bible was of divine origin.[32]

Matthew Maury (1806-1873) was a fine Christian who is credited with founding the science of hydrography. Hydrography refers to the study of oceans, lakes, and rivers. He charted the currents and winds of the Atlantic for the U.S. Navy. Maury also taught meteorology at the Virginia Military Institute.[33] He was probably the first to recognize the relationships between the circulation of air over the earth and the circulation of ocean currents. This knowledge enabled him to chart the best routes for ocean vessels. Maury was inspired to undertake such a career by reading Psalm 8:8 which refers to "the paths of the seas." He reasoned that if the Bible referred to such paths, then they must really exist and should be studied.[34]

Sir James Simpson (1811-1870) was a Christian physician who was born in Scotland. He taught obstetric medicine at Edinburgh University and is said to be one of the founders of gynecology. He also discovered chloroform in 1847 which was a major contribution to the field of anesthesiology. Sir James is also remembered for his writings on fetal pathology and his studies of leprosy in Scotland. Sir James wrote a gospel tract which contains the following passage; "But again I looked and saw Jesus, my substitute, scourged in my stead and dying on the cross for me. I looked and cried and was forgiven. And it seems to be my duty to tell you of the Saviour, to see if you will not also look and live."[35]

James Joule (1818-1889) was another pioneer of science who was a creationist.[36] This English physicist did research in heat, electricity, and thermodynamics. His most important accomplishment, which came in 1840, was finding the value of the "mechanical equivalent of heat" constant. Joule's constant allowed heat energy to be quantitatively converted into mechanical energy and vice versa. This constant made the mathematical expression of the law of conservation of energy possible.[37]

Gregor Mendel (1822-1884) was an Austrian monk whose classic experiments revealed a basic flaw in Darwinism. He conducted careful experiments with peas, between 1854 and 1863 which established the basic principles of heredity. According to Henry Morris, these studies "established the basic stability of the created kinds of plants and animals." Mendel's studies proved, among other things, that characteristics are inherited as separate units and do not mix. He never accepted Darwinism which was rapidly gaining in popularity at that time.[38]

Louis Pasteur (1822-1895) was another Christian scientist who revealed a flaw in Darwinism. Pasteur struck a serious blow to evolution by disproving the concept of spontaneous generation. He is also responsible for the "germ theory" as it relates to disease. Consequently, Pasteur ranks as one of the greatest figures in the history of medicine. He also made important contributions in the fields of physics and chemistry. Pasteur taught chemistry at the University of Strasbourg.

Louis Pasteur

Henri Fabre (1823-1915) was another French Christian who was an outstanding scientist. Fabre was a friend of Pasteur and was an active opponent of the doctrine of evolution. He is credited with establishing the field of entomology because of his extensive studies of insects. Fabre authored many children's books about science. These were used as textbooks for public schools in France until some (so-called) intellectuals raised objections to his many references to God as Creator. Fabre was recipient of numerous honors for his scientific accomplishments.[39]

William Thompson, Lord Kelvin (1824-1907) was a Christian who was a great physical scientist and inventor. He had 70 patents. Lord Kelvin opposed both Darwinism and uniformitarianism. He was born William Thompson in Belfast, Ireland and matriculated at the University of Glasgow at the age of ten. In 1846, after studying at Cambridge, he became professor of natural philosophy at Glasgow and served in that position for 53 years. He is remembered for a large number of achievements in mathematics and physics, as well as his many practical inventions. In 1852 Thompson presented the second law of thermodynamics. According to Burchfield, the second law "reconciled Carnot's theory of the inefficiency of heat engines and Joule's principle of conservation by distinguishing the total energy of a system from the energy available for useful work." Thompson was knighted in 1866 as a result of his supervision for the laying of the first trans-Atlantic telegraph cable. He was president of the Royal Society from 1890 to 1895. He received nearly every scientific honor imaginable, including 21 honorary doctorates.[40]

Joseph Lister (1827-1912) was an English surgeon of Quaker background who once wrote that he was a "believer in the fundamental doctrines of Christianity." His major contribution was the improvement of surgery by using chemical disinfectants. Lister began by using carbolic acid but moved on to other chemicals when he found that carbolic acid damaged tissue. This was an enormous improvement that greatly enhanced the survivability of surgical procedures. Lister founded the Institute of Preventative Medicine and also served as president of the Royal Society.[41]

Sir Henry Rawlinson (1810-1895) was a dedicated Christian and avid student of the Bible who was born in Chadlington, England. He was one of the great archaeologists and spent part of his life in the English military stationed in India. Rawlinson is best remembered for deciphering the cuneiform Behistun inscriptions of Darius the Great. These inscriptions are found on the face of a towering scarp in western Iran known as the Behistun Rock. They are written in the ancient languages of Assyrian, Elamitic, and Old Persian. It has been said, "this accomplishment opened the way to a real understanding of the ancient history of the Near and Middle East." He was appointed British consul to Baghdad in 1843, and after being knighted in 1855, Rawlinson served as crown director of the British East India Company.

Sir Joseph Henry Gilbert (1817-1901) was an agricultural chemist who was a member of the Royal Society of science. He had a strong faith in the Bible and opposed Darwinism. His research led to nitrogen and super-phosphate fertilizers that were used to enhance the growth of crops. In 1843, Gilbert became the co-director of the first agricultural experiment station[42] at Rothamsted, England. His work at Rothamsted proved (among many other things) that if chemical fertilizers are used in conjunction with proper soil management, an acceptable level of soil fertility can be maintained for many years.[43]

Thomas Anderson (1819-1874) was an outstanding Christian chemist from Scotland, who was also a member of the prestigious Royal Society. Anderson is credited with discovering the organic base pyridine among other contributions. He was selected to be editor of the *Edinburgh New Philosophical Journal*. He also signed the *Scientist's Declaration of 1864*, which was a document that "affirmed their confidence in the scientific integrity of the Holy Scriptures." This *Declaration* was a response to the growing influence of Darwinism and was signed by hundreds of scientists including 86 members of the Royal Society.[44]

Edward H. Maunder (1851-1928) was a Christian astronomer and was the foremost authority on solar astronomy of his day. He founded and served as president of the British Astronomical Association. He authored a book about Biblical references to astronomy in which he discussed the Bible's accuracy and credibility on astronomical topics. For six years Anderson served as secretary of the much revered Victoria Institute. This British institution is dedicated to the defense of the Christian faith.[45]

William Mitchell Ramsay (1851-1939) was a professor at Aberdeen University and Oxford University. According to Josh McDowell, he "was one of the greatest archaeologists ever to have lived." Ramsay did considerable archeological research in Asia Minor. He was converted to Biblical Christianity when his archeological discoveries confirmed the historical statements of the book of Acts (written by Luke) After decades of study, Ramsay said, "Luke is a historian of the first rank," which meant that the historical events of Acts were entirely accurate.[46] Ramsay wrote over 20 books, most of which presented archeological evidence in support of the New Testament.

A.H. Sayce (1845-1933) was a Christian archeologist and philologist from England. He did archeological and linguistic studies that helped vindicate some historical passages of the Old Testament. Sayce was recognized as an expert on the Assyrians and Hittites. As with Ramsay, the archeological studies of Sayce led him to trust Christ for salvation. He was author of over 25 books which included Early History of the Hebrews, published in 1879. In 1890 during a trip to Egypt, Sayce found the Constitution of Athens which was written by Aristotle. He brought this document, which had been given up as lost, back to England for the British Museum.

John Ambrose Fleming (1849-1945) has been called the father of modern electronics and developed the first electron tube. He taught electrical engineering for decades at the University of London. Fleming was one of Maxwell's students at Cambridge and later served as a consultant to Marconi and Thomas Edison. He made numerous contributions to the fields of radio, television and electronics. He was the son of a Congregational minister. Fleming authored an important book against evolution and founded the Evolution Protest Movement.[47]

George Washington Carver (1864-1943) was born a slave and later worked his way through college where he prepared for a career in science. He finished a master's degree at Iowa State in 1896. He was eventually recognized as the foremost authority on peanuts, sweet potatoes and their related products. Carver is credited with developing over 300 products as a result of his research with the peanut. This great agricultural chemist served on the faculty at the Tuskegee Institute. Carver was a sincere Christian with a strong testimony. He was awarded the Roosevelt medal in 1939.

Douglas Dewar (1875-1957) earned a degree in natural science from Cambridge and became a naturalist and ornithologist in India. He wrote over 20 books concerning the birds and history of India. Dewar was an evolutionist until about age 50 and even wrote books defending evolution. When he became a creationist Christian, Dewar then began writing books about the scientific foundation of creationism. He defended the creationist position in debates with British evolutionists such as J.B.S. Haldane, Joseph McCabe, and H.S. Shelton. In 1932, Dewar helped to found the Evolution Protest Movement in London.[48]

Walter Wilson (1881-1969) was an amazingly versatile Christian physician who was born in Aurora, Indiana. He earned the MD degree from the

University of Kansas and originally practiced medicine in Webb City, Missouri. Early in Wilson's career, his father-in-law became very ill. His father-in-law had been a tentmaker and Wilson felt that he should try to keep the family business going. So Dr. Wilson became a part time doctor and a full time tentmaker. His scientific background enabled him to develop a process for waterproofing and camouflaging tents, in response to a request from General Pershing. Buffalo Bill was also said to have been a customer of their tentmaking business. However, Wilson didn't just make tents the rest of his life and went on to establish and pastor the Central Bible Church in Kansas City. He also opened the Kansas City Bible Institute in 1932, which is now called Calvary Bible College. Dr. Wilson wrote numerous books, pamphlets, and articles on both religious and medical themes. Besides all these accomplishments, he was considered to have been a pioneer of the radio ministry.[49] John R. Rice wrote, "Dr. Wilson was a kindly Christian physician who knew how to drive home to the heart how Christians ought to live. He was also one of America's great and best-loved Bible teachers."[50]

Wernher von Braun (1912-1977) was a director of NASA who was recognized as being one of the foremost space scientists in the world. He earned a PhD from the University of Berlin and worked for Germany as a rocket engineer during WWII. He is credited with developing the V-2 rocket. After that war, he came to this country and became a naturalized citizen in 1955. von Braun was an active Lutheran and said, "I find it as difficult to understand a scientist who does not acknowledge the presence of a superior rationality behind the existence of the universe as it is to comprehend a theologian who would deny the advances of science."

L. Merson Davies (1890-1960) was an outstanding Christian geologist and paleontologist who was active in the Evolution Protest Movement. Davies earned both a PhD and a DSc in geology and frequently wrote significant articles for geological journals in England. He debated J.B.S. Haldane and other evolutionists. Davies authored a book in which he defended the Bible's scientific accuracy.[51] He believed that II Peter 3:3-6 referred to the doctrine of uniformitarianism. Dr. Davies wrote that (up until the time of Darwin and Lyell), "belief in Deluge of Noah was axiomatic, not only in the Church itself, (both Catholic and Protestant) but in the scientific world as well. And yet the Bible stood committed to the prophecy that, in what it calls the 'last days,' a very different philosophy would be found in the ascendent; a philosophy which would lead men to regard belief in the Flood with disfavor, and treat it as disapproved, declaring that 'All things continue as from the beginning of the creation' (II Peter 3:3-6)."[52]

Depiction of the Ark on Mount Ararat

CHAPTER 5

THE ARK

The Technical Feasibility of Noah's Ark

The Bible says in Genesis 6:12-14, "All flesh had corrupted his way upon the earth. And God said unto Noah, The end of all flesh is come before me; for the earth is filled with violence through them; and, behold, I will destroy them with the earth. Make thee an ark of gopher wood; rooms shalt thou make in the ark, and shalt pitch it within and without with pitch." Because of the corruption and violence of the antediluvian world, God has decided to destroy it and Noah is told to build the Ark.

According to John Woodmorappe, "attacks on the credibility of the Ark account go back to classical antiquity." He further wrote, "Apelles, a disciple of the heretic Marcion is listed as one of the first critics of the Ark. He asserted that the Ark was barely large enough for four elephants! The early church fathers responded to some of these criticisms."[1] The first part of this chapter will attempt to answer the following question. Is the concept of a huge wooden boat built to house thousands of animals for a year-long ocean voyage feasible?

There are some definite specifications that the Bible gives us concerning Noah's Ark. The Bible says in Genesis 6:15-16, "This is the fashion which thou shalt make it of: The length of the ark shall be three hundred cubits, the breadth of it fifty cubits, and the height of it thirty cubits. A window shalt thou make to the ark, and in a cubit shalt thou finish it above; and the door of the ark shalt thou set in the side thereof: with the lower, second, and third stories shalt thou make it." Dr. Morris and Dr. Whitcomb calculate the size and capacity of Noah's ark based on a value of 17.5 inches for the cubit. They write that the "Ark was 437.5 feet long, 72.92 feet wide, and 43.75 feet high. Since it had three decks (Genesis 6:16), it had a total deck area of approximately 95,700 square feet (equivalent to slightly more than 20 standard college basketball courts), and its total volume was 1,396,000 cubic feet."[2] Concerning the Ark's window, Dr. Harold Sightler wrote that it "probably extended all the way around the boat at the top. Just one window and that was right at the top of the boat a cubit, eighteen inches wide and the window proved to be the only source of light and the only source of ventilation."[3] It is likely that the pitch mentioned in verse 14 was a resinous compound[4] or might have been a type of asphalt substance. The gopher wood was probably either cyprus or cedar.[5]

Noah's Ark was probably the largest ocean vessel ever built, until the late nineteenth century when huge metal ships began to be constructed. The length to breadth ratio, according to Scripture, was 6 to 1. It is interesting that the huge tankers of modern times have ratios of about 7 to 1. There were Danish barges that were closely modeled after the Ark. These barges were found to be highly seaworthy and very difficult to capsize. In 1844, I.K. Brunel designed a ship called the *Great Britain* which had dimensions of 322 feet x 51 feet x 32 1/2 feet. These ratios are also very similar to those of the Ark, so it is clear that the idea of a sea-going vessel such as Noah's Ark is really quite plausible.[6] Concerning Noah's Ark, Dr. Morris said, "It can be shown hydrodynamically that a gigantic box of such dimensions would be exceedingly stable, almost impossible to capsize. Even in a sea of gigantic waves, the ark could be tilted through any angle up to just short of 90% and would thereafter right itself again."[7] The Ark was really more like a barge than a ship. It was rectangular and probably flat on the bottom. There is no mention in the Bible of any mast or rudder. The fact that Noah's Ark had no rudder meant that God could providentially take the Ark where ocean conditions would be safest at any given time.[8]

Noah's Ark had an enormous capacity. Dr. Whitcomb wrote, "For the sake of realism, imagine waiting at a railroad crossing while 10 freight trains, each pulling 52 box cars, move slowly by, one after another. That is how much space was available in the Ark, for its capacity was equivalent to 520 modern railroad stock cars."[9] Now a double-decked stock car can carry about 240 sheep. So if we assume that a sheep was the average size for animals on the Ark,[10] we can multiply 240 x 520 to get the total number of sheep-sized animals that could conceivably fit into the Ark. This gives us a figure of nearly 125,000 animals. But how many animals did Noah's Ark really carry?

The Bible says in Genesis 7:2, "Every clean beast thou shalt take to thee by sevens, the male and his female: and of beasts that are not clean by two, the male and his female." Dr. Morris wrote, "There are (according to leading taxonomist Ernest Mayr) less than 20,000 species of land animals (mammals, reptiles, birds, amphibians) living today with a much smaller number of extinct species known from the fossils. At the most, therefore, the ark would only have to carry, say, 80,000 animals." This figure is based on the assumption, obviously, that the "kind" of Genesis 6:20 corresponds to the species. However, Dr. Morris also points out that "the 'Biblical kind' is very likely much broader than the species in most cases, so there would have been no problem at all, in accommodating all the world's land based animals in the ark."[11]

In spite of the fact that creationists have demonstrated that the "kind" of the Bible is broader than the species, "anti-creationists" (such as Moore) still say the Ark had to carry every species. The Bible says in Genesis 1:24, "God said, Let the earth bring forth the living creature after his kind, cattle, and creeping thing, and beast of the earth after his kind: and it was so." This verse indicates that the term "kind" mainly has to do with reproduction.

Mr. Woodmorappe wrote, "There is a wealth of evidence that, at minimum, the created kind is broader than the species of conventional taxonomy." It is a mistake to assume that the biblical "kind" is equivalent to the term species. The term syngameon refers to "the most inclusive unit of interbreeding among plants and animals." In most cases this syngameon is broader than the species and can even be broader than the genus. After showing that the "kind" corresponds more closely to the genus than the species, Woodmorappe arrives at a figure of about 16,000 total animals on the Ark.[12] The great majority of the animals on the Ark were probably small, perhaps about the size of a rat (about 100 grams).[13]

We find in Genesis 6:20-22, "Two of every sort shall come unto thee, to keep them alive. And take thou unto thee of all food that is eaten, and thou shalt gather it to thee; and it shall be for food for thee and for them. Thus did Noah; according to all that God commanded him, so did he." Concerning this passage, in the *KJV Parallel Bible Commentary* we find that God "advised Noah to lay in store in the ark all the food necessary for his family and the animals. The last verse of this chapter is most remarkable. Although God has explained that He will destroy all those on the earth, He has not yet explained to Noah how this will be accomplished."[14] Dr. Sightler wrote, "No doubt he was indeed the object of scorn, criticism, and ridicule and yet Noah believed God and prepared an ark."[15]

Let's take a look at stored food on the Ark. Hay, for example, can be stored for some time if properly cured. Woodmorappe writes, "Year-old hay (the duration of the Ark voyage) is commonly fed to as diverse as horses and elephants. As is the case with hay, various dry feeds keep for a long time if their moisture is sufficiently low. The know-how for preserving feedstuffs for durations of at least three years was known in Biblical times (Leviticus 25:21-22)."

Of course the preservation of grain requires drying and also the prevention of moisture seeping back into the grain. Critics have said that wet conditions inside the Ark would have ruined these feedstuffs. However, water-tight containers of various kinds could have been used to protect seeds and feedstuffs from wet conditions, if such conditions did exist. This is not a procedure requiring high technology.[16]

Other critics of the Ark account often say that an enormous amount of hay would have been required to feed all the large mammals on Noah's Ark. These detractors claim that this great bulk of hay could not have fit on the Ark. However, this criticism is based on the false assumption that the dietary needs of large mammals can only be supplied by hay. This criticism is also based on the assumption that hay taken aboard Noah's Ark had to be in its usual low-density condition. In fact, for the most part, grain could have been substituted for hay. Obviously, grain in storage requires far less space than hay. An 80% grain/20% hay diet, for the great majority of the large mammals on the Ark, is not a problem from an animal science viewpoint.[17]

It is known that during lengthy journeys in Biblical times the hoofed mammals (also called ungulates) carried a type of concentrated grain called provender which served as their feed. This provender was far less bulky than hay. Pigs and other non-ruminants do not require fibrous feed and so can be given a diet of strictly grain. Ruminants such as cows and sheep have been raised on all-grain diets on many occasions. For those ungulates, such as horses, that cannot subsist on an all grain diet, fibrous feeds such as oats have been substituted for hay. Furthermore, Noah may have been able to compress the hay on the Ark. It would not have required a very high level of technology to double the density of that hay. So the criticism concerning the bulk of hay on Noah's Ark is simply without foundation.[18]

Mr. Woodmorappe attempted to calculate the amount of food taken on the Ark. He wrote, "This is done by expressing it in terms of dry matter intake which scales closely according to animal body mass. I first calculated the dry matter intake of all the animals on the Ark and then converted it to the intake of actual possible foods. The total dry-matter intake on the Ark comes out to 1990 tons." He accounts for the fact that warm-blooded animals have a much greater food requirement than do cold-blooded animals. The consumption of feed by the large mammals far exceeded that of the rest of the animals combined. There were probably a wide variety of feedstuffs taken on Noah's Ark. This mostly had to do with preventing nutritional deficiencies. If these feedstuffs had a moisture content of 20% on average, the total mass of food on the Ark was about 2,500 tons. This figure is based on 16,000 total animals which generally represented each genus of land animals.[19]

So how much space did all this food take up? Woodmorappe addresses this issue in his book where he wrote, "Were all of the requisite 1990 tons of dry matter food on the Ark in the form of settled hay, just over half the volume of the Ark would have been occupied by it." However, he goes on to say that if "an average of 80% rate of substitution, the food required for the year long voyage would have occupied less than a sixth of the Ark's volume."[20]

Mr. Woodmorappe also attempts to calculate the total amount of potable water that had to be carried on the Ark. He assumes that this water had to be stored on the Ark even though Noah may have had the means to collect rainwater. Woodmorappe wrote, "Growing animals need drinking water in a proportion greater than a simple fraction of their respective adult weight. For this reason, the same Juvenile Factors (JF) are used to calculate water consumption of the young large mammals as when calculating animal food intake." He concludes that about 4.1 million liters of water was taken. This amount of water would have occupied 9.4% of the Ark's total volume assuming a value of 18 inches for the cubit.[21]

But how much of the Ark's volume was occupied by the animals themselves? The two modern examples of animal confinement most closely resembling the Ark would be that of animals in the laboratory and animals in the "factory farm" In the factory farm situation there may be as many as 100,000 animals that are housed together with just a few people to care for them. Woodmorappe used the standards of these two types of animal housing as a basis for calculating the floor space requirements of the animals on the Ark.

Noah's Ark was really an emergency shelter for the animals. Even though their surroundings may have seemed strange and confining, these animals would have gradually become accustomed to them. The conditioning of animals to accept stressful situations is called habituation. The fact that the animals boarded the Ark a week before the beginning of the Flood (Genesis 7:9, 10) probably had to do with habituation. Woodmorappe gives the example of rhinos which "are fed in traveling crates for two weeks prior to travel." These rhinos are more manageable and generally have an easier time during travel. He estimates that the boarding of the 16,000 animals would have taken up to about five hours.[22]

Mr. Woodmorappe wrote, "Claims that captive animals require large areas of housing in order to survive are largely untrue." There must have been a great deal of motion associated with the voyage of the Ark. There may have been, and probably were, periods of relative calm but let's remember that a global cataclysm was taking place. Spacious enclosures would have allowed drastic movements by those animals. This would have been very dangerous to them. He goes on to say, "It is the smallness of animal enclosures on the Ark (table 3) which, in addition to the previously discussed habituation of animals to stress, must have been the main factor protecting the animals from injury." It is also true that nervous animals can be more easily handled when they are in small enclosures. According to Woodmorappe's calculations, no more than about 47% of the Ark's floor space was needed for animal housing.[23] So the animals (47% of floor space), plus their food (15.4% of the Ark's volume), plus the potable water (9.4% of volume) could have fit on the Ark with plenty of room to spare.

Mr. Woodmorappe's calculations were based on an 18 inch cubit. However, the cubit may have been longer than 18 inches. Dr. Rice wrote in his commentary on Genesis that "a cubit was originally intended to be the measure of a man's forearm to the tip of his fingers. Ordinarily that will be 18 to 27 inches. The cubit varied because there was no standard measurement put away in Washington, DC, by which all the yardsticks were made as now."[24] If the cubit, referred to in Genesis 6:15, was actually 27 inches then the length of the Ark would have been slightly more than an eighth of a mile.

Skeptics of the Ark account have argued that the animal waste on Noah's Ark would have been an insurmountable problem. However, this isn't necessarily true. Much of this waste could have been consumed by earthworms. When manure is consumed by an earthworm, it goes through the worm's gut and comes out as a dry casting This casting is odorless.[25] Research suggests that the consumption of organic waste by earthworms reduces levels of pathogenic bacteria.[26] The earthworms could have been placed in pits below cages and in gutters under stalls. Constructing floors, in the animal housing area, with a slight to moderate slope would have facilitated this process. The rest of the animal waste could have been collected periodically and dumped overboard.[27] Woodmorappe believes that God's directions for taking "two of every sort" (Gen. 6:20) only referred to vertebrates. This view is based on a study of the Hebrew text.[28] So there would have been no restriction to taking great numbers of earthworms on the Ark.

Likewise, large numbers of fireflies could have been used as a source of illumination within the Ark. Woodmorappe wrote, "Even in fairly recent times, Chinese and Japanese who could not afford to buy fuel for oil lamps, or candles, would gather fireflies into a bag to make a lantern by which to read." Besides fireflies, there is a luminous bacteria which might have been used on the Ark for illumination.[29]

Another objection raised by detractors of the Ark account is that eight people supposedly could not have cared for the thousands of animals on the Ark. Mr. Woodmorappe addresses this issue in chapter 8 of his book. In that chapter he wrote, "Based on actual manpower studies (and with no miraculous assistance nor animal dormancy), eight people could have definitely taken care of tens of thousands of animals." Although it is true that some animals, in certain situations, may require extensive care, this isn't really typical. He points out that the "American farm horse in the early 20th century, by contrast, required an average of only seventeen minutes of total labor daily." At the present time, horses need much less attention, since modern stables have devices that reduce the required daily labor.

One laboratory technician has been known to care for 5,900 laboratory animals. Other examples are one man caring for 3,840 pigs and a single stockman responsible for 30,000 laying hens. Even when people are working in a non-mechanized situation, they can still individually care for hundreds, perhaps thousands of animals (e.g. 842 pigs, 1000 ducks, 2000 pigeons). Noah and his family may have benefitted from rudimentary labor saving devices which were not unknown in ancient times. Mr. Woodmorappe also wrote that people in ancient Rome "knew about mass feeding and watering of both domestic and wild birds, using feeding and drinking gutters that were filled with water with pipes."

Self feeding and the piping of water into troughs are two of the labor saving techniques that could have been used on the Ark. Noah's Ark could have been constructed so that bins were attached to the floor. The appropriate feed, depending on what animals were in close proximity, might have been placed into these bins before the Flood. These bins could have been periodically uncovered for self-feedings. Woodmorappe wrote, "Plimer (1994, p. 128) makes the farcical claim that the Ark crew must have been limited to the use of buckets to move water within the Ark. For his information, the history of plumbing goes back to early antiquity. The ancient Minoans and Egyptians had extensive plumbing."[30]

The Spiritual Implications of the Sacrificial Animals on the Ark

Some animals were taken on the Ark by sevens. The Bible says in Genesis 7:2, "Of every clean beast thou shalt take to thee by sevens, the male and his female." These clean beasts were acceptable for sacrifice. Dr. Ben David Lew said, "When Noah came out of the Ark, he sacrificed seven clean animals to God."[31] In Genesis 8:20 we read, "And Noah builded an altar unto the Lord; and took of every clean beast, and of every clean fowl, and offered burnt offerings on the altar." Dr. Rice wrote, "Abel had offered animal sacrifices, and so Noah offers animal sacrifices too. This is before the ceremonial law, but even then Abel and Noah understood that the sacrifice represented Christ, the Lamb of God who would come and die."[32]

When you read in the Bible of "a little lamb slain and his blood spilt," wrote Dr. Curtis Hutson, "God is saying, 'I am going to send a Redeemer.'

Every time you see little doves killed and their blood spilt, God is saying, 'I am going to send a Redeemer.'"[33] The Bible says in Ephesians 1:7 that "we have redemption through His blood, the forgiveness of sins, according to the riches of His grace." Concerning this verse, Oliver B. Greene said, "By every animal, innocent and without blemish, slain in sacrifice, according to the direction of Jehovah God, God was saying through the death of that animal, 'My Lamb, My Son is going to die!'"[34]

These animals that were to be sacrificed had to be without any defects. The Bible says in Leviticus 22:20, 21, "...whatsoever hath a blemish, that shall ye not offer: for it shall not be acceptable for you. And whosoever offereth a sacrifice of peace offerings unto the Lord to accomplish his vow, or a freewill offering in beeves or sheep, it shall be perfect to be accepted: there shall be no blemish therein." Then over in the New Testament, in I Peter 1: 18, 19, the Bible says, "Ye know that ye were not redeemed with corruptible things, as silver and gold, from the vain conversation received by tradition from your fathers. But with the precious blood of Christ, as of a lamb without blemish and without spot." There is a continuity between the Old Testament and the New Testament. The concept that Jesus was the sacrificial Lamb of God was proclaimed by John the Baptist in John 1:29.

The Bible says in that verse, "The next day John seeth Jesus coming unto him, and saith, Behold the Lamb of God, which taketh away the sin of the world." The Jews of that time must have known what John the Baptist was talking about. Their Temple was being reconstructed by Herod and the Temple service was taking place, including sacrificial offerings. They knew that the Temple sacrifices represented the coming Saviour, or Messiah. So when John the Baptist cried "Behold the Lamb of God," in reference to Jesus, it should have immediately registered in their minds that Jesus was the Messiah (i.e. anointed one, or Christ). People in the Old Testament times were saved by their faith in the coming Messiah. The great fundamentalist, John R. Rice, wrote, "Dimly it may be they saw, but by faith they did see that God would provide a Sacrifice."[35]

From John 1:29 we know that the sacrificial offering of the Lamb of God was all about sin. This Sacrifice was the crucifixion of Jesus Christ on Calvary about two thousand years ago. We may not like to admit it, but we are all sinners. We read in Romans 3:23 that "all have sinned and come short of the glory of God." Likewise, the Bible says in Romans 3: 10 that "...There is none righteous, no, not one." We find the same idea expressed in Isaiah 5 3:6 where the Bible says, "All we like sheep have gone astray; we have turned every one to his own way." In Ecclesiastes 7:20 we read, "there is not a just man upon earth, that doeth good, and sinneth not."

Sin must be paid for because it carries a penalty, or debt. In Romans 6:23 the Bible says, "For the wages of sin is death." This death is both physical and spiritual. Ezekiel 18:4 says, "The soul that sinneth, it shall die." Then over in Revelation 20:14 we find "Death and hell were cast into the lake of fire. This is the second death." Dr. Hutson put it very succinctly when he wrote that "if we pay what we owe as sinners, we must spend an eternity in the lake of fire (Hell)."

However, Jesus has already paid our sin penalty, or debt, when He died on the cross. The Bible says in Romans 5:8, "But God commendeth his love toward us, in that, while we were yet sinners, Christ died for us." Dr. Hutson wrote in his gospel tract, *How to Know You are Going to Heaven*, that the Scripture "teaches that God took all of our sin and placed it on Christ. While Christ was bearing all of our sins in His own body, God punished him in our place to pay the debt we owe."

The Bible says in John 3:16, "For God so loved the world, that he gave his only begotten Son, that whosoever believeth in him should not perish, but have everlasting life." The word 'believe' means to trust, to depend on, to rely on."[36]

In Romans 5:9 we see, "Much more then, being now justified by his blood, we shall be saved from wrath through him" Now, the realization of the connection between our own personal sins and the blood of Christ is absolutely vital. The Bible says in I John 1:7 that "the blood of Jesus Christ his Son cleanseth us from all sin." In large measure, the cleansing blood of Christ is the foundation of all Scripture. Concerning I John 1:7, Charles Spurgeon wrote, "Only in truth and holiness can we have fellowship with God, and to render this possible to such sinful creatures as we are, the precious blood of Jesus must purge us from sin."[37] The Bible says in Revelation 1:5 that Christ "washed us from our sins in his own blood."

Please notice that in Romans 3:24, 25 the Bible says, "Being justified freely by his grace through the redemption that is in Christ Jesus: Whom God hath set forth to be a propitiation through faith in his blood." The word propitiation means an atoning sacrifice. The word atonement refers to the reconciliation of God and man through the sufferings and death of Jesus. A reconciliation is necessary because God, who is infinitely righteous, can't just let sin go unpunished. The perfection, or purity, of God is referred to as holiness. In Habakkuk 1:13 the Bible says that God is "of purer eyes than to behold evil and canst not look on iniquity." The justice of a holy God must be satisfied. In other words, your sin debt must be paid. We can pay it ourselves and spend eternity in Hell or trust Christ who has already paid everyone's sin debt on the cross. It is true that your sin debt has already been paid. But it won't be applied to your account (so to speak) unless you place a conscious trust in the atoning blood of the sacrificial Lamb of God (Romans 3:25) who was Jesus Christ. If it weren't for the blood of Christ, there would be no hope of reconciliation. The Bible says in the ninth chapter of the book of Hebrews, "Almost all things are by the law purged with blood; and without shedding of blood is no remission."

Even when some people grasp the significance of the relationship between the blood of Christ and their own personal sins, they still don't think that faith in the Lamb of God is adequate for salvation. They just can't believe that simply trusting the Lamb of God, by calling on Him for mercy, cleansing from sin and an everlasting home in Heaven, is really sufficient. They try to mix faith with works to gain eternal life. They may think that it is necessary to live by the golden rule, be baptized, join a church, keep the ten commandments, do benevolent deeds, or something else to make it to Heaven. They just don't believe that trust in the Lamb of God is sufficient.

However, the Bible says in Revelation 5:12, "Worthy is the Lamb that was slain." The word "worthy" in this verse means "sufficient." So you might just as well say "sufficient is the Lamb." The Bible says in Titus 3:5 "Not by works of righteousness which we have done, but according to his mercy he saved us, by the washing of regeneration, and renewing of the Holy Ghost." Living a good life is not sufficient but the Lamb is sufficient. Likewise, in Ephesians 2:8, 9 we find, "By grace are ye saved through faith, and that not of yourselves: it is the gift of God: Not of works, lest any man should boast." These verses clearly establish the fact that redemption is an act of God, not an act of man. Some people have the erroneous notion that the Lamb of God is their parachute, so-to-speak, and that living a good life is their reserve parachute. They just don't think that Christ is entirely reliable as far as their salvation is concerned. So they must do so many good deeds as a kind of back-up. But the Bible tells us in Revelation 5:12 that the Lamb is sufficient to save you.[38]

In Exodus 12:5 the Bible says, "Your lamb shall be without blemish, a male of the first year." Then over in I Peter 1:19 the Bible says that redemption is by the "precious blood of Christ, as of a lamb without blemish" In I Cor. 5:7 we read, "For even Christ our Passover is sacrificed for us." It would be just as correct to say that Christ is our Passover lamb. These verses clearly show that the Passover lamb of Exodus 12 is meant to represent Jesus, the Lamb of God.

But there is more proof. The Bible says in John 19:13, 14, "When Pilate therefore heard that saying, he brought Jesus forth, and sat down in the judgment seat in a place that is called the Pavement, but in the Hebrew, Gabbatha. And it was the preparation of the Passover." Then, a little further on in verse 31, the Bible says, "the Jews therefore, because it was the preparation, that the bodies should not remain upon the cross on the sabbath day, (for that sabbath day was an high day) besought Pilate that their legs might be broken, and that they might be taken away." Concerning this verse, Dr. Rice wrote, "It is the sabbath of Exodus 12:16, the day of the Passover supper and the first day of the feast of unleavened bread, a day of 'an holy convocation'." The Jews wanted the bodies of Jesus and the thieves taken down before sundown when the Sabbath would begin.[39] According to Dr. Ben David Lew, "the word Easter occurs in the Scriptures. There is a verse in the King James Version of the New Testament where the word is found, but there the word should be translated Passover as it is in the original text (Acts 12:4)."[40] It is important to understand that when Jesus died on the cross, he did so as that last, great Passover sacrifice. Joe Henry Hankins wrote, "The very day was set fifteen hundred years in advance. That night when the Passover lamb was slain in Egypt, the date was set upon which the Son of God should die. When the hour approached, the Pharisees said, 'Not on the day of Passover, lest a tumult arise.' But God had said fifteen hundred years before, that it must be on the Passover—the hour when our Passover should be slain."[41]

What can wash away my sin?
Nothing but the blood of Jesus;
What can make me whole again?
Nothing but the blood of Jesus.
Oh precious is the flow
That makes me white as snow;
No other fount I know,
Nothing but the blood of Jesus.[42]

Another animal that was acceptable for sacrifice was the goat. Goats were used extensively for sacrifices and also for feasts. In Old Testament times, goats were sacrificed in observance of the solemn Day of Atonement. Dr. Ben David Lew wrote that the "word *Yom Kippur* itself meaning the Day of Atonement, comes from a Hebrew word *Kapper*, 'to cover'. According to this meaning, God covers the sins of His people by the blood of the sacrifice." He also writes, "This expresses perfectly the idea of substitutionary death and propitiation. It points to the heart of the Gospel: Christ died for our sins,' I Corinthians 15:3."[43]

Two goats were needed for the Day of Atonement. The Bible says in Leviticus 16:8-10, "Aaron shall cast lots upon the two goats; one lot for the Lord, and the other lot for the scapegoat. And Aaron shall bring the goat upon which the Lord's lot fell, and offer him for a sin offering. But the goat, on which the lot fell to be the scapegoat, shall be presented alive before the Lord, to make an atonement with him, and to let him go for a scapegoat into the wilderness." Concerning Leviticus 16:10, Spurgeon wrote, "Thus our great substitute bears away the sins of his people into oblivion."[44] This same thought is expressed in Jeremiah 50:20 where we read, "In those days, and in that time, saith the Lord, the iniquity of Israel shall be sought for, and there shall be none; and the sins of Judah, and they shall not be found." Just as the scapegoat is sent into the wilderness, when a person trusts Christ for salvation God sends their sins far away. God removes the convert's sins so far away, "They shall not be found."

In order for a person to be saved, the sin question must be addressed. No one can get saved without, first of all, recognizing that he's a sinner. That first step is absolutely vital. Salvation is really all about addressing the issue of sin. That's why baptism, church membership, observing the sacraments, living a good life or speaking in tongues don't have any power to save a person from Hell. None of these actions really address the issue of sin. The Bible says in Isaiah 64:6, "We are all as an unclean thing, and all our righteousnesses are as filthy rags; and we all do fade as a leaf; and our iniquities, like the wind, have taken us away."

Because of our sinful, depraved nature we can't even begin to measure up to God's perfect standard. Yet multitudes of religious people, including most religious Jews, are trying to do this very thing. Ben David Lew wrote, "Judaism could best be summed up as man's attempt to justify himself by his own effort, without the atonement made by a Saviour. That cannot be done. There can be no substitute for Christ, our Atonement and Propitiator."[45] The cross of Christ is the only hope for the Jew, as well as the Gentile.

No discussion of Jesus, the Lamb of God, would be complete without mentioning the resurrection. What an awesome miracle that was! Dr. Sightler said that when he was in an ancient history class, the professor said, "The greatest and the best established fact in all of ancient history [that would include the Caesars, the early days of the Roman empire the Grecian empire and Alexander the Great] is the resurrection of Jesus Christ." He also said that the "best established fact in all history is that on the third day Jesus came out of the grave!"[46]

The Jewish historian Josephus, writing near the end of the first century A.D., wrote, "There was a man about this time Jesus, a wise man, if it be lawful to call him a man; for he was a doer of wonderful works, a teacher of such men as receive the truth with pleasure. He drew over to him many Jews, and also many of the Greeks. This man was the Christ. And when Pilate had condemned him to the cross, upon his impeachment by the principal man among us, those who had loved from the first did not forsake him for he appeared to them alive on the third day." These facts were also reported by Tertullian in the second century.

The Gospel of Luke contains an account of the crucifixion, resurrection, and ascension of Jesus. It is remarkable that Luke uses the pronoun "we" a number of times in that book. This indicates that Luke was a participant in at least some of the events that he wrote about. Luke was well acquainted with the facts of the matter regarding the life and death of Christ. In Luke 24:1-3, we read, "Now upon the first day of the week, very early in the morning, they came unto the sepulchre, bringing the spices which they had prepared, and certain others with them. And they found the stone rolled away from the sepulchre. And they entered in, and found not the body of the Lord Jesus."[47]

R.A. Torrey said, "The resurrection of Jesus from the dead is the best proven fact of history; and the absolutely certain resurrection of the Lord Jesus Christ proves to a demonstration that He is the Christ, the Son of God."[48] Dr. Torrey was a graduate of Yale and later studied at the University of Hamburg in Germany. He was chosen by D.L. Moody to be superintendent of the Moody Bible Institute. Dr. Torrey also helped establish the Bible Institute of Los Angeles and held a leadership position there. He also pastored the Church of the Open Door in Los Angeles. Dr. Torrey was a highly respected pastor, evangelist, and educator with tremendous credibility.

Just as the death of Jesus, the Lamb of God, was miraculous, so also was his birth. The Bible says in Isaiah 7:14, "The Lord himself shall give you a sign; Behold, a virgin shall conceive and bear a son and shall call his name Immanuel." The prophecy is fulfilled in Matthew 1:23, "A virgin shall be with child, and shall bring forth a son, and they shall call his name Emmanuel, which, being interpreted is, God with us." Even though liberal theologians have denied the validity of the virgin birth, there are good reasons to take Isaiah 7:14 literally.

Much of the controversy hinges on the meaning of the word virgin. In Isaiah 7:14 the word "virgin" was translated from the Hebrew word "almah". "Almah" refers to a woman who isn't married, in the Hebrew language. There are six other places in the Old Testament where "almah" appears. In every case, "almah" means an unmarried woman. Genesis 24:43 is a good example. The Bible says in this verse, "Behold, I stand by the well of water; and it shall come to pass that when the virgin (*almah*) cometh forth to draw water, and I say to her, Give me, I pray thee, a little water of thy pitcher to drink." Concerning this verse, Ben David Lew wrote, "These words were spoken by Eliezer, Abraham's servant, and that virgin referred to was Rebekah, who was certainly not a married woman." Other examples of *almah* referring to an unmarried woman can be found in Exodus 2:7, 8, Psalm 68:25, Song of Solomon 1:3 and 6:8, and also in Proverbs 30:19. Dr. Ben-Lew wrote, "In every case where the word *almah* is used in the Hebrew Bible, it invariably means a virgin and nothing else."

The Septuagint is believed to have been the first translation of the Hebrew Old Testament into the Greek in about 270 B.C. This translation, from the original Hebrew, was done by Jews. They translated *almah* to *parthenos* which means "virgin" in the Greek language. This demonstrates that Jews of that era used the word *almah* to refer to a virgin. In Luke 1:27 the Bible says, "The virgin's name was Mary." The word for virgin in the original Greek is *parthenos*, exactly the same word that the Jews used for *almah* in the Septuagint.[49]

McDowell points out, in his book *Evidence That Demands a Verdict*, "In Isaiah 7:14 the birth is said to be a 'sign' from the 'Lord Himself'. This is certainly unique in that this could be no ordinary birth. Thus, we can see that the doctrine of the virgin birth presented in the Gospel is in accord with earlier Scripture teachings."[50] To deny the virgin-birth is to undermine the deity of Christ.

The salvation offered by Jesus, the virgin born Lamb of God, is available to everyone. The Bible says in Romans 10: 12, 13, "There is no difference between the Jew and the Greek: for the same Lord over all is rich unto all that call upon him. For whosoever shall call upon the name of the Lord shall be saved." Then over in Revelation 22:17 we read, "Let him that is athirst come. And whosoever will, let him take the water of life freely."

Have you ever trusted Jesus, the Lamb of God, for salvation? Are you sure? If you haven't, why not thank Jesus for dying on the cross to pay your sin debt? Then call on Him right now for mercy, forgiveness, and an everlasting home in Heaven. As Lester Roloff said, "You are helpless. You are dead in trespasses and sin. Not the church nor Mother nor Dad nor the priest...nor the baptismal waters nor good works can save you. You can't do a thing about it. Then just make a confession; just admit that Jesus Christ died for your sins according to the Scriptures. Just take the Bible plan that says, 'Now is the accepted time; behold, now is the day of salvation' (II Cor. 6:2). Then trust Him and say 'Come into my heart, Lord Jesus, and save me.'"[51]

CHAPTER 6

THE FLOOD

The Bible says in Genesis 7:11, 12, "In the six hundredth year of Noah's life, in the second month, the seventeenth day of the month, the same day were all the fountains of the great deep broken up, and the windows of heaven were opened. And the rain was upon the earth forty days and forty nights." So in these verses we find that the two main sources of the Flood's waters are mentioned. Let's look at what is meant by the term "fountains of the great deep" and the other term "windows of heaven."

According to Whitcomb and Morris, "the phrase t^e hôm rabbâh ('great deep') points back to the t^e hôm of Genesis 1:2 and refers to the oceanic depths and underground reservoirs of the antediluvian world. Presumably, then, the ocean basins were fractured and uplifted sufficiently to pour waters over the continents."[1] In the antediluvian (pre-Flood) world there must have been enormous quantities of water trapped in reservoirs deep within the earth. We know this because the Bible speaks of rivers in Genesis 2:10-14. In Genesis 2:5 we find, "The Lord God had not caused it to rain upon the earth", so these antediluvian rivers could not have been produced by the run-off of rainfall. Dr. Morris has written that these rivers "emerged through controlled fountains or springs, evidently from deep-seated sources in or below the earth's crust." Proverbs 8:24 refers to "fountains abounding with water" which were probably the very same thing as the fountains of Genesis 7:11. Morris also writes that "such subterranean reservoirs were apparently all interconnected with each other, as well as with the surface seas into which the rivers drained, so that the entire complex constituted one 'great deep'." He believes that subterranean heat provided the energy to recycle and repressurize these waters. So the reference to the fountains in Genesis 7:11 really speaks of a point in time when the waters in these subterranean reservoirs burst out of confinement and escaped to the surface.

Dr. Morris reasons that since this fracturing of the fountains is mentioned first in Genesis 7:11, then this action must have set off other events. Genesis 7:11 indicates that these subterranean conduits became fractured on the same day. He has written, "The most likely cause would seem to have been a rapid buildup and surge of intense pressure throughout the underground system, and this in turn would presumably require a rapid rise in temperature throughout the system." It would be virtually impossible to determine exactly what triggered such a rise in temperature. However, there are a number of possibilities which include seismic activity, volcanic activity, and even nuclear reactions within the earth. That volcanic activity was one factor seems certain.[2]

Whitcomb and Morris write in reference to volcanic activity during the Flood, "great quantities of liquids, perhaps liquid rocks or magmas, as well as water, probably steam, had been confined under great pressure below the surface rock structure of the earth since the time of its formation and that this mass now burst forth through great fountains, probably both on the lands and under the seas."[3] The Flood was accompanied by stupendous earth movements, volcanic activity, and tremendous climatic and geomorphic changes.[4]

So what are the "windows of heaven" in Genesis 7:11? This is a term that refers to the same thing as the "waters which were above the firmament" in Genesis 1:7. Creationists believe that these waters were actually a vast water vapor canopy that surrounded the entire earth. This vapor canopy had a greenhouse effect and caused global weather patterns to be uniformly mild and rather tropical. Since this body of water was in the vapor state, it was transparent. This canopy must have contained an enormous quantity of water.[5] The very fact that Genesis 7:12 speaks of rain lasting forty days points to the existence of such a vapor canopy. Under present conditions, all the atmospheric water would be precipitated in a rather short time and would only average about two inches of rainfall worldwide. Whitcomb and Morris address this issue in *Genesis Flood* where they write that the "normal hydrologic cycle would, therefore, have been incapable of supplying the tremendous amounts of rain the Bible record describes. The implication seems to be that the antediluvian climatology and meteorology were much different from the present. There seems to have been an atmospheric source of water of an entirely different type and order of magnitude than now exists."[6]

Dr. Joseph Dillow describes the vapor canopy model in chapter five of his book *The Waters Above* where he writes, "This pre-Flood atmosphere contained sufficient water vapor to sustain a 40-day-and-night rainfall of about 0.5 inches per hour as discussed in chapter two. This amounts to about 40 feet of water and, hence, 2.18 atmospheres of atmospheric pressure on the pre-Flood earth." He goes on to say in that same chapter, "The cause of the condensation of the canopy was apparently the action of volcanoes in hurling volcanic ash into the atmosphere, providing condensation nuclei for the rain to form on."[7] Tremendous amounts of volcanic dust blown into the atmosphere, along with towering jets of water (released from underground reservoirs), and general atmospheric turbulence would have combined in effect to penetrate the vapor canopy. Once the condensation nuclei were within the canopy it would have soon started to rain.[8]

The condensation and precipitation of this global canopy of water vapor profoundly affected the world's topography and climate. According to Whitcomb and Morris, "the pre-diluvian topography was completely changed with great mountain chains and deep basins now replacing the formerly gentle and more uniform topography. Removal of the protective canopy around the earth permitted development of extreme, latitudinal variations of temperature, with resulting great air movements and established climatic zones. Removal of the canopy also permitted the earth's atmosphere to be penetrated by much larger amounts of radiation of various types."[9]

This vapor canopy would have created a uniformly mild and, more or less, subtropical climate over the whole earth. Dillow writes, "In such an environment a greenhouse effect would have resulted and the temperature differentials would be negligible, resulting in only minor wind movements, and hence, no rain. Instead, the earth was watered by a 'mist'. During the cool of the day, the temperature would drop by a few degrees and the water vapor in the atmosphere would condense out as a mist."[10]

The fossil record strongly suggests that a mild and moderate climate once existed around the world, even at the higher latitudes. The fossil remains of a very lush vegetation indicates that there was either a tropical or subtropical climate from what was supposedly Cambrian time up to the Miocene epoch. The coal formations of the Pennsylvanian strata can really only be explained by warm and moist conditions that would produce luxuriant vegetative growth all year.[11] According to Whitcomb, "trillions of tons of vegetation, much of it perfectly preserved, even to leaves and flowers, have been buried in all parts of the world, including Antarctica, in the form of coal. Each foot of coal represents many feet of compressed plant remains, and some coal seams are as much as thirty or forty feet in thickness."[12] There are other indications that there must have been a moderate world climate in the past. For example, remnants of coral reefs that could only have been formed by tropical sea creatures are found in Arctic waters. These remains, that normally occur in warm waters, are found near the poles. Fossilized tropical animals have been found in Greenland and other, northern latitudes.[13]

Prior to the Flood, the earth's crust was in a state of isostatic balance. However, this equilibrium was destroyed by a number of factors. These factors included the enormous quantities of water and volcanic material that came to the surface, the huge amounts of sediments created by and transported by the Flood, and the profound alterations in the geomorphic characteristics of the antediluvian world. The term isostasy refers to the principle of equal weight. This means that, at a given level beneath the earth's surface, the pressure exerted by overlying materials must be the same all around the world. This condition is a state of equilibrium. It just follows that areas of low topography will be more dense than areas of higher topography.[14]

When weight in the crust, or outer layer, of the earth is shifted, the crust and the magma will adjust to achieve equilibrium. The magma, which lies just under the crust, is actually molten rock. When weight is transferred from one area to another, the area of the crust which acquires the weight (obviously) becomes heavier. Heavier segments of the crust will sink and this pressure will squeeze the underlying magma out laterally. This magma can actually move very slowly. When the magma reaches an area of the crust that is lighter, the pressure of the magma will push this lighter area up. The movement of this creeping magma will continue until the total weight of magma plus crust, down to a certain depth, is approximately equal between adjacent regions.

The Earth's crust seems to have contracted in the past to form the great mountain ranges. This contraction of the crust was similar to the shrunken skin on a dried apple. The wrinkles on the apple's skin are analogous to mountain ranges.[15] The immense amounts of lava and gasses which were ejected from deep within the Earth during the Flood caused the molten interior of the Earth to actually shrink. This shrinkage must have seriously weakened the support for the crust. Lava is magma that has reached the surface. As the molten interior of the Earth shrank, due to the enormous loss of heat and magma, the rigid crust would have resisted contraction. However, the crust at some point would have collapsed under its own weight, in the most seriously weakened areas.

The antediluvian isostatic balance was completely disrupted. So, as the Flood waters retreated and dried, a new isostatic balance was being established. Significant isostatic adjustments may have continued for a long time.[16] Ham, Snelling, and Wieland wrote, "As the mountains rose and the valleys sank, the waters would rapidly drain off the newly emerging land surfaces. Such rapid movement of large volumes of water would cause erosion, and so it is not hard to envisage the rapid carving out of many of the landscape features that we see on the earth today, including, for instance, places like the Grand Canyon of the USA." Many river valleys are far larger than the rivers that flow through them. These valleys seem to have been carved out in the past by a much larger quantity of water than the rivers that now flow through them have. This phenomenon is consistent with the concept that they were eroded by the waters of the Noahic Flood that drained off of the surface of the land that emerged from the Flood. Most of these waters ran off into the deep ocean troughs that were formed at the end of the Flood.[17] The re-emergence of lands had to be accompanied by a tremendous geologic upheaval to create deep basins to hold these waters. Areas that are now the continents, including great mountain ranges, were uplifted as a consequence of isostatic adjustments.[18]

Dr. Rice wrote, "In Job also there seems to be some memory of the great convulsions that took place on the earth following the flood. Inland seas and lakes broke out and washed out rivers and flowed to the ocean." Job 28:4 says, "The flood breaketh out from the inhabitant; even the waters forgotten of the foot: they are dried up, they are gone away from men." Job 28:9-11 says, "He putteth forth his hand upon the rock: he overturneth the mountains by the roots. He cutteth out rivers among the rocks; and his eye seeth every precious thing. He bindeth the floods from overflowing; and the thing that is hid bringeth he forth to light."[19]

There is another factor besides isostasy that hastened the retreat of the Flood's waters. With the collapse of the vapor canopy the "modern cycle" for precipitation was established. This cycle involved evaporation and condensation which is followed by either rainfall or snowfall. Whitcomb and Morris write that, by this cycle, "large amounts of water were removed from the oceans and stored in the polar regions in the form of great ice caps, which in some instances are believed by glacial geologists to have attained the immense size of continental ice sheets." Vast amounts of ice and snow were produced as the result of the collapse of the water vapor canopy. The glaciers must have continued to grow until they reached a point at which summer thawing offset winter accumulation. The total quantity of water that was held in these glaciers was actually much greater after the Noahic Flood than now. There is evidence that sea level was about 400 feet lower at one time than it is now. This lower level seems to have been the result of an Ice Age when enormous amounts of water were trapped in the glaciers and polar ice caps, as ice.[20]

That there was an abrupt and permanent drop in polar temperatures is predicted by the canopy model. When the canopy was precipitated the moderate global climate would have been drastically altered. On the basis of this model, we could predict that an abundance of plants and animals would have abruptly disappeared. [21]

There is really only evidence for one ice age. This best correlates with the Pleistocene ice age, which was the most recent ice age according to evolutionists. Creationists believe that the Ice Age started at about the same time as the Flood and continued for another several hundred years. Evidence of an Ice Age can still be seen in the great ice caps in alpine glaciers, in glacial sediments, and in landforms caused by glaciers. It seems reasonable that this Ice Age must have been relatively recent for these effects to still be observable. During this period a vast sheet of ice covered much of North America and actually carved out the Great Lakes. Northern Europe was also covered by a similar sheet of ice which extended from Scandinavia to Italy. Another huge expanse of ice covered an area from Antarctica to the southeast portion of the Australian mainland.[22] Whitcomb has written, "The end of the ice age is much more recent than was once speculated, and there is much evidence now available that there was only one great glaciation, not four."[23]

Dr. Larry Vardiman has written that "the Ice Age in the Biblical time frame probably lasted from shortly after the Flood to about the time of Abraham. The Ice Age probably affected the people of the Bible with a cooler, wetter climate and more grasslands in the Tigris-Euphrates Valley, Palestine, and North Africa. These regions are known for their desert climate today but evidence points to more vegetation in the past." One of the factors that contributed to the Ice Age was the warming of the oceans. This was due to the enormous amount of heat that was released by volcanic activity. This volcanic activity occurred during and for an extended period after the Flood. Much of the volcanic activity would have occurred under the oceans and would have directly warmed the water. The oceans may have been 20° C warmer than at present.

This higher oceanic temperature would have resulted in a number of atmospheric conditions. Vardiman lists four of these conditions. The warmer water would have evaporated more easily and this would have led to much more precipitation, especially snow at higher latitudes. Second, the abrupt temperature differential between continents and the oceans would have resulted in higher precipitation and strong winds in coastal areas. Third, unprecedented growth of microscopic organisms in the oceans would have resulted in high levels of carbon dioxide (CO_2) being absorbed by these organisms. The lower CO_2 levels in the atmosphere would have resulted in lower air temperatures. Fourth, the difference in temperatures between the water and the air would have enhanced convection, which is the vertical movement of air. Increased convection would also have resulted in higher amounts of precipitation.

The rapid growth of microorganisms was due to warmer water, food added to the oceans by the Flood, and by the presence of fewer predators after the Flood. The combination of these three factors created a situation that allowed the microorganisms to double their biomass in a short period of time. These organisms were actually very tiny plants that used carbon dioxide for photosyntheses. As the ocean waters began to cool there would have been a decrease in biomass, and less carbon dioxide would have been tied up. The estimate of water temperature is calculated on the basis of the composition of *foraminifera* shells. These shells are taken from sea floor sediments. *Foraminifera* incorporate variable amounts of oxygen isotopes into their shells depending on water temperature.[24]

There is a relationship between carbon dioxide levels and air temperatures. CO_2 is one of the atmospheric gasses that allows solar radiation to reach the earth's surface in the form of light. The CO_2 doesn't reduce the amount of light that reaches the surface in any way. The light is absorbed by solids or liquids on the surface. The absorbed light heats the liquid or solid. Then this heat is emitted into the atmosphere in the form of infrared radiation. CO_2 is not transparent to infrared and will collect this type of heat and hold it in the air.[25] So as the oceans cooled, and with the resulting decrease in microorganism biomass, atmospheric levels of CO_2 would have increased. As these CO_2 levels were built up, air temperatures would have begun to rise. Then as air temperatures rose the ice would have begun to recede.

Genesis 10:25 says, "And unto Eber were born two sons: the name of one was Peleg; for in his days was the earth divided; and his brother's name was Joktan. " So we find here that the earth was divided during the lifetime of Peleg. The earth of this verse is translated from the Hebrew word erets. Erets is also the word used for earth in Genesis 1:1. John R. Rice wrote, "It is the physical earth that was divided, not the people, at the confusion of tongues." As previously mentioned, the Ice Age which occurred right after the Flood caused sea level to be much lower than at present. The significance of this fact is that, because of the low water level, "land bridges" would have connected the continents. So after the Flood, men and animals were able to walk to all the continents. This means that the world's human and animal populations could have originated on the Ark.

It has been calculated that if about 12,000,000 cubic miles of water could be returned to the North and South poles, sea level would fall by about 500 feet. If this happened large areas now covered with water would be dry land, including the continental shelves. Land bridges would connect Alaska with Russia, and Asia with Australia. North America and Europe may also have been connected by a land bridge. In fact, such a bridge possibly connected Africa with South America. The significance of Genesis 10:25 is that sometime during the life of Peleg, sea level rose enough to cover these land bridges. So that the earth was divided simply means that the continents were separated by water.[26]

The Ussher Chronology places the time of the Flood at about 2348 B.C. According to the genealogy given in Genesis 11, Peleg was born 101 years after the Flood and lived 209 years. So somewhere between 2247 B.C. and 2038 B.C. sea level rose enough to cover the land bridges with water. It would be reasonable to say that this event occurred in the middle years of Peleg. Otherwise it might have been more accurate to say that the earth was divided in the days of Eber, Peleg's father, or in the days of Reu, Peleg's son. So these land bridges were probably actually covered with water somewhere between 2200 B.C. and 2100 B.C.[27]

Frank L. Marsh has written, "As regards the Bering Strait, there is no doubt that a land connection once existed between Asia and North America.

With the strait closed, the cold waters of the Arctic would have been pre-vented from coming south, and the Japan Current would have curved around the coast line farther north than today. The washing of those shores by the warm waters of this current would have produced a dry-land route that even tropical forms could have used." Marsh also points out that the island group known as the East Indies would have been the bridge that the marsu-pials crossed to reach Australia when sea level was lower than at present. If these two gaps were connected, then a "dry-land path" would exist be-tween the habitable continents.[28]

Tremendous amounts of volcanic ash from continuing eruptions after the Noahic Flood would probably have been in the atmosphere. This air-borne volcanic material would have reflected considerable solar radiation back into space. The loss of solar radiation would have resulted in a large drop in air temperatures. This volcanic material was probably a factor for hundreds of years after the Noahic Flood because of widespread volcanic activity. The presence of airborne volcanic dust kept air temperatures low for a very long time. Enormous amounts of volcanic rocks have, in fact, been found mixed in with Tertiary and Pleistocene sediments which cre-ationists believe were deposited soon after the Flood.[29]

The book of Job seems to have a number of references to an Ice Age. According to the Ussher Chronology the events of Job happened in 1520 B.C. which is about the same time that Moses spent in Midian. This would be about 825 years after the Flood. Job seems to have lived in the waning years of the Ice Age while the climate in Bible lands was wetter and cooler than at present. Exodus 3:8 indicates that there was abundant vegetation, at least in some areas, in Bible lands at that time. Job lived in the land of Uz which was a region located just to the south of Edom. The country called Edom was just to the south of Israel.[30]

As Dr. Morris has written, "there are more references to cold, snow, ice, and frost in Job than in any other book of the Bible." For example, in Job 37:9-10 the Bible says that "out of the south cometh the whirlwind and cold out of the north. By the breath of God frost is given: and the breadth of the waters is straitened." Job 38:22 says, "Hast thou entered into the trea-sure of the snow? or hast thou seen the treasures of the hail." In Job 38:29 we read, "Out of whose womb came the ice? and the hoary frost of heaven, who hath gendered it?" This last verse suggests a sudden and perhaps sur-prising appearance of ice and otherwise cold conditions.[31]

Now if volcanic ash really was the agent that caused the precipitation of the canopy, then it must have been washed down with the torrential downpour. It seems reasonable to suspect, then, that some of this ash would still be imbedded in glaciers that were formed as a consequence of the plung-ing temperatures, when the vapor canopy began to collapse. Dr. Dillow has written, "Army cold regions geologist Anthony Gow took 7,100 feet of core samples from nine Antarctic glaciers. He found over 2,000 individual volcanic ash falls imbedded with the ice. Gow suggests that these volcanic eruptions may have contributed to the cooling of the Antarctic atmosphere and brought about the Ice Age." Large amounts of volcanic material have also been found in the frozen tundra muck mixed in with the frozen remains of animals and plants. For this to have occurred, the muck must have been soft and more or less, in a state of turbulence. Concerning these phenomena, Dillow has written, "This again indicates a sudden cooling along with volcanic activity, precisely the prediction of the canopy model."[32]

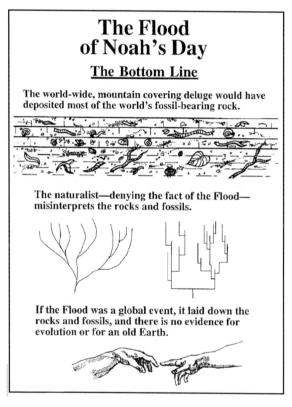

The Flood
of Noah's Day
The Bottom Line

The world-wide, mountain covering deluge would have deposited most of the world's fossil-bearing rock.

The naturalist—denying the fact of the Flood—misinterprets the rocks and fossils.

If the Flood was a global event, it laid down the rocks and fossils, and there is no evidence for evolution or for an old Earth.

There is another observation that helps to confirm the vapor canopy model. This is the fact that water vapor has been detected in the stratosphere. It seems reasonable that if such a vapor-canopy did exist, then a small remnant of this canopy might still be in the higher atmosphere. In fact, $H^+ (H_2O)_6$ ions have been found at an altitude of 85 km. This corresponds to the D layer of the ionosphere, The evolutionists do not really have an explanation for how water vapor ions could reach the ionosphere but this phenomena is entirely consistent with the canopy model.[33]

A global flood is also consistent with the vapor canopy model. Dr. Dillow writes, "Without a doubt the collapse by condensation of such a vast vapor canopy would result in untold devastation and deluge all over the planet. For 40 feet of water to pour from the heavens over all the earth for a period of 40 days and nights (0.5 inches per hour) would unleash a fantastic flood catastrophe." If such a global catastrophe really did occur, there must be evidence of that fact. Dillow suggests that there are a number of evidences to suggest that a great global food, such as that described in Genesis chapters 7 and 8, did occur. The first evidence is that of "polystrate fossils". A common example would be a fossilized tree which extends through several layers of strata which are supposed to have been deposited over millions of years. These trees prove that the sediments were deposited fairly rapidly. Otherwise the tree would have decayed a very long time before the sediments were deposited, and would not have been fossilized at all.[34] Dr. Parker has written, "Palm trees washed out in vegetation mats after a tropical storm, may float upright for a while, and they could be entombed in that upright position if burial occurred quickly enough." Polystrates occur frequently in coal formations.[35]

Another evidence that a worldwide flood occurred is the absence of worldwide unconformities. Morris has written in chapter three of his book *Science and the Bible*, "An unconformity is a surface between two geological formations, an interface on which erosion has taken place between the two periods of sediment deposition represented by the strata below and above that interface. Such an unconformity may be recognized by a sudden change in the characteristics of the strata." This can be a change in types of fossils, the classification of rock, or the angle of strata. He goes on to say in that same chapter that an "unconformity surface typically is formed when an uplift raises the local sediments above the water level, so that deposition on the surface stops and erosion begins. Thus, an unconformity means a time gap in the deposition process." Even though localized unconformities have been observed, there is not any unconformity of a worldwide extent, except at the base of the geologic column. So there are no global gaps in deposition. Therefore, the whole concept of the eras, periods, and epochs in the geologic column must be erroneous.[36]

Each stratum was formed as a result of a particular group of hydraulic factors that cannot exist in combination for any length of time. This means that each stratum must have been deposited fairly rapidly. A stratum which overlays another with no sign of unconformity must have been deposited immediately after the preceding stratum. Since unconformities have not been found to separate any two strata, except on a localized basis, successive strata must have been deposited continuously. So if each stratum was deposited rapidly and if successive strata were deposited continuously, then the complete geologic column must have been deposited rapidly and, more or less, continuously. Furthermore, there is seldom a clear boundary between strata.[37]

Another evidence of a global flood is that of the fossils themselves. Dr. Dillow considers this to be "one of the most obvious evidences of a flood." The fossilized remains of plants and animals are normally found in sedimentary rock. In order for the fossilization process to occur, there must be a rapid burial to prevent decay. The dead plant or animal must also remain undisturbed while this process takes place. The fossils occur in virtually every stratum and this is true on a global basis. So it just follows that there must have been a world-wide catastrophe which produced the rapid burial of billions of organisms.[38] The multitudes of fossils that are preserved in the fossil record couldn't have been produced in a manner that is now observable. Fossilization is actually a very rare occurrence. The world's fossil record can only be the result of a global, hydraulic catastrophe.[39]

A scriptural proof of the global extent of the Flood is the fact that God told Noah to build the Ark. Dr. Whitcomb writes, "The whole procedure of constructing such a vessel, involving over one hundred years of planning and toiling, simply to escape a local flood, can hardly be described as anything but utterly foolish and unnecessary. How much more sensible it would have been for God simply to have warned Noah of the coming destruction in plenty of time for him to move to an area that would not have been affected by the Flood."[40]

All of the world's mountains, as they existed during the Flood, were under fifteen cubits of water (at least). Fifteen cubits was half the height of the Ark and probably represented the portion of Ark that was underwater. So the Ark could float over all the mountains. This would have included the mountains of Ararat. The highest mountain of this range is about 17,000 feet.[41] Moses was inspired by God to write in Genesis 7:24, "The waters prevailed upon the earth an hundred and fifty days." Concerning this verse, the *KJV Parallel Bible Commentary* says, "Water, which seeks its own level, could not continue to rise for 150 days in a local Mesopotamian valley flood. It is inconceivable that the Great Flood was anything else but a universal judgment of God upon a universally wicked society."[42]

Another proof of a global Flood is found in Matthew 24:37-39 that says "But as the days of Noe were, so shall also the coming of the Son of man be. For as in the days that were before the flood they were eating and drinking, marrying and giving in marriage, until the day that Noe entered into the ark. And knew not until the flood came, and took them all away; so shall also the coming of the Son of man be." In this passage the event of the Flood is presented in an analogy to the event of Christ's return. That the Rapture, mentioned in verses 40-42, will be global in scope is a clear teaching of the Bible. So if the Noahic Flood was local in extent these two sets of verses would not be analogous at all. In II Peter 2:5 we find that God "spared not the old world, but saved Noah the eighth person, a preacher of righteousness, bringing in the flood upon the world of the ungodly." This verse indicates that only Noah and the other seven in the Ark survived the Flood. This could not have been true if the Flood was less than global in extent. We find in Hebrews 11:7, "Noah, being warned of God of things not seen as yet, moved with fear, prepared an ark to the saving of his house; by the which he condemned the world, and became heir of the righteousness which is by faith." This is another verse that is only sensible in the context of a global Flood.

Scriptural evidence to support a universal Flood is also found in Genesis 17:19-20 where the Bible says, "The waters prevailed exceedingly upon the earth: and all the high hills that were under the whole heaven were covered. Fifteen cubits upward did the waters prevail: and the mountains were covered." Chapter 7 of Genesis reveals that all of the mountains were covered by the end of the first 40 days and that this depth was maintained for another 110 days. Because of the influence of gravity on levels of water, the Noahic Flood would have had to be global if such a depth of water was maintained for even a week.

Referring to the seventh chapter of Genesis, Morris writes, "The wording of the entire record, both here and throughout Genesis 6-9, could not be improved on, if the intention of the writer was to describe a universal Flood; as a description of a river overflow, it is completely misleading and exaggerated, to say the least."[43]

Cloud Vapor

CHAPTER 7

THE FOSSIL RECORD

Sedimentary Rocks and Uniformitarian Interpretation

Sedimentary rocks are formed from sediments which have accumulated during periods of deposition. Such rocks are formed when vast amounts of sediment accumulate, or settle, in a low area. Over time, the bottom portion of this sedimentary material becomes rock. This transformation is largely due to the pressure from above. Sand that is deposited will become sandstone. Mud will become shale. Gravel, that is deposited, will be pressed and cemented into conglomerate rock.

Sedimentary rocks are composed of consolidated, or solidified, sediments that were deposited in layers, called strata. The unconsolidated sediments are converted to solid rock by a number of processes. These processes include cementation, compaction, and desiccation.[1] Cementation involves the filling of the voids between sediments by chemicals that will crystallize. Calcium carbonate ($CaCo_3$ silica (Sio_2) and dolomite ($CaMg(Co_3)_2)_1$ can serve as cements to bring about the hardening of sedimentary material. Compaction, which reduces the volume of pore space, generally increases in effect with depth. Desiccation refers to the removal of moisture from sediment.[2]

The weathering and erosion of preexisting rock gradually produces sediment. The erosion of soil is an even greater source of sediment. A number of agents will move this sediment from its original location to its site of deposition. These agents include running water, moving ice, wind, and gravity. Running water, along with gravity, are the most important agents. Whitcomb and Morris wrote, "Almost all of the sedimentary rocks of the earth, which are the ones containing fossils and from which the supposed geologic history of the earth has been largely deduced, have been laid down by moving waters. This statement is so obvious and so universally accepted that it needs neither proof nor elaboration."[3] They go on to say, "The deposition occurs, of course, when the running water containing the sediments enters a quiescent or less rapidly moving body of water, the lowered velocity resulting in a dropping out of part or all of its load of moving sediment."[4]

Evolutionists believe that the waterborne sediments that became the sedimentary rocks were deposited very slowly, in accordance with the concept of uniformitarianism Uniformitarianism has been defined as "the present is the key to the past."[5] Referring to uniformitarianism, Dr. Gish wrote, "According to this interpretation of earth's history, existing physical processes, acting essentially at present rates, are sufficient to account for all geological formations. As originally formulated by James Hutton and Charles Lyell, any appeal to catastrophes for the explanation of geologic phenomena is rejected."[6] The evolutionists do not reject a catastrophic interpretation of earth history because of scientific observation. The reason that they reject such an interpretation is because of their philosophical bias against the Bible, especially the great Flood of Genesis which would have had a profound geological impact on the Earth.

The evolutionists classify sedimentary rocks according to the types of fossils that they contain. Index fossils are believed to have been deposited during a definite and relatively short span of time. Relatively short means in comparison to the vast extent of assumed earth history. These index fossils are used to identify and date portions of the world's fossiliferous strata. For example, trilobites are one type of index fossil. Sedimentary rocks that contain fossilized trilobites are said to be Cambrian rocks. These Cambrian rocks are alleged to represent the first period of the Paleozoic Era.

The so-called Cambrian rocks are assumed to have been followed by a number of other distinct sedimentary deposits in a chronological order. Each of these types of fossiliferous deposits is assumed to have been deposited over millions of years. All these different types of sedimentary deposits arranged in an imagined time sequence are usually referred to as the geological column. This time sequence is based on the assumption of the validity of organic evolution. An examination of this geological column/time scale reveals how closely the ideas of Charles Darwin and Charles Lyell fit together.[7]

Because of the fact that no scientist actually saw the fossiliferous strata being deposited, the geological column really represents a set of assumptions. Furthermore, since the evolution of one species into another species has not actually been observed either, this geological column is really just a set of assumptions based on an underlying assumption. According to Huse, "the fact that modern historical geology is based on the assumption of evolutionary biology is a blatant case of circular reasoning. The only basis for placing rock formations in chronological order is their fossil content, especially index fossils. The only justification for assigning fossils to specific time periods in that chronology is the assumed evolutionary progression of life. In turn, the only basis for biological evolution is the fossil record so constructed." The geological column actually proves nothing about either evolution or geologic time.

Dr. Huse goes on to say, "It is important to realize that nowhere in the world does the geologic column actually occur. It exists only in the minds of evolutionary geologists."[8] The idea that all the various systems of strata in the geological column can be found at any point below the earth's surface, on a global basis, is a misconception. Dr. Steven Austin wrote, "Approximately 77% of the earth's surface area on land and under the sea has seven or more (70% or more) of the strata systems missing beneath; 94% of the earth's surface has three or more systems missing beneath; and an estimated 99.6% has at least one missing system." Dr. Austin, who earned the Ph.D. degree in geology at Penn State, was referring to the ten systems of strata that are said to contain the proof of evolution. These systems include the Cambrian, Ordovician, Silurian, Devonian, Carboniferous, Permian, Triassic, Jurassic, Cretaceous, and Tertiary. Furthermore, these systems are not always in the order that the geologic column requires. There are hundreds of places where the actual order of systems of strata doesn't match the theoretical sequence of the geologic column.[9]

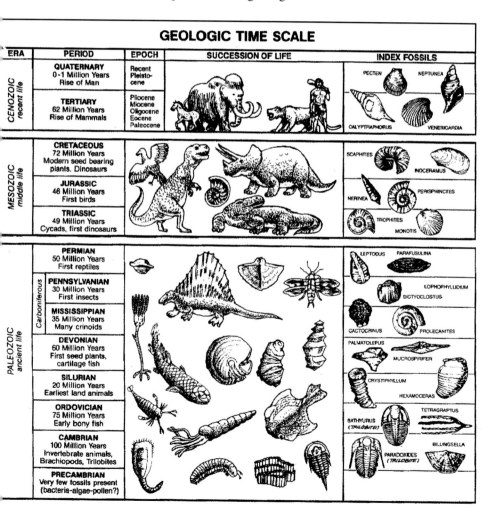

The great number of situations where strata and fossils are found in the wrong places, present an enormous problem to geologists who advocate uniformitarianism For example, there are many cases where formations that are supposed to be younger (i.e. deposited more recently) are actually beneath rock formations that are supposed to be older. Such cases are found very frequently in mountainous areas and are said to be due to thrust-faulting. The concept of thrust-faulting, basically, is that enormous segments of rock strata are supposed to break loose from their original position and then slide considerable distances over adjacent rocks. The younger layers of these sedimentary rocks, which have moved, are subsequently eroded. According to Whitcomb and Morris, "The concept is that great segments of rock strata have somehow separated from their roots and made to slide far over adjacent regions. Subsequent erosion then modifies the transported 'nappe' so that the young strata on top are removed, leaving only the older strata superposed on the stationary younger rocks beneath."

There are a number of different versions of this thrust-faulting concept but all involve a vertical movement of rock strata followed by a horizontal movement of the same rock strata. The vertical movement is said to be preceded by the development of a sloping or vertical fault that separates rocks. This fault, or break, is said to occur in rock strata that was once continuous. Besides the mechanical problems, there is usually very little, if any, evidence that these catastrophic movements of the huge blocks of rock strata actually happened.[10] It is something of an irony that the evolutionist geologists must flee to catastrophism to account for anomalies in the real-world geologic column.

Dr. Whitcomb wrote in chapter three of *The World That Perished* that an "example of the failure of the uniformity principle to account for significant features of the earth's crust is the 'thrust fault' hypothesis, as applied to such areas as the Heart Mountain Thrust of Wyoming and the Lewis Overthrust of Montana and Alberta. The latter formation which includes the Glacier National Park area, is 350 miles long, and has an inferred horizontal displacement of at least thirty-five or forty miles, in spite of the fact that the supposed fault plane dips at an angle of only three degrees. The black rocks on the upper half of the mountain are Precambrian while the lighter-colored rocks below are Cretaceous." Whitcomb goes on to say in that same chapter, "On the basis of known friction coefficients for sliding blocks, so much shearing stress would be developed in a large block that the material itself would fail in compression and, therefore, could not be transported as a coherent block at all." Therefore, the Cretaceous rocks in that vicinity must have been deposited before the Precambrian rocks.[11]

Another example of sedimentary strata that is not in the right sequence, according to the standard geologic column, is the Heart Mountain Thrust.

According to Whitcomb and Morris, "this supposed thrust occupies roughly a triangular area, 30 miles wide by 60 miles long, with its apex at the northeast corner of Yellowstone Park It consists of about 50 separate blocks of Paleozoic strata (Ordovician, Devonian, and Mississippian) resting essentially horizontally and conformably on Eocene beds, some 250,000,000 years younger!" They also wrote that there are "many pictures of the 'fault-line', all of them looking for all the world like any other normal contact between chronologically deposited strata. An even more mysterious factor is that there appear to be no source beds from which the thrust blocks could have broken off."[12]

It is common for fossils to be found out of place, as far as the expected evolutionary sequence is concerned. These out of place fossils are called displacements. Concerning displaced fossils, Whitcomb and Morris wrote that 'if it is supposed to be older than the containing bed, it can be said to have been redeposited from an earlier eroded deposit or to indicate the survival of its particular species longer than had been previously believed. If it is supposed to be younger than its stratum, it can be again explained as due to the reworking and mixing of two originally distinct deposits or else as showing that the animal dates from earlier antiquity than previously thought."[13]

Other explanations that the uniformitarian geologists offer, to explain deposited layers that are out of the expected theoretical order, are overturning and erosion. Overturning is said to be a consequence of catastrophic physical deformation. When layers of rock strata are overturned they are alleged to be bent over into a sort of a loop. At some point this loop is imagined to break off, leaving a section of rock strata in reverse order (i.e. upside down).[14] Sometimes beds of rock strata are missing, as far as the theoretical geologic column is concerned. In such cases, the missing beds are said to have been deposited but were later eroded away.[15]

Creationists and evolutionists are in sharp disagreement over the rate of deposition of the sediments that became the world's sedimentary rocks. The evolutionists believe that this deposition was very, very gradual. They claim that the sedimentary rock strata were deposited over a time period of millions upon millions of years. However, creationists believe that the deposition of the sediments, which became the rock strata, was fairly rapid. They also believe that the deposition of this enormous amount of sediment was the result of a global catastrophe, described in the seventh and eighth chapters of Genesis.

Dr. Huse wrote, "Creationists maintain that uniformitarian principles simply cannot account for most of the major geologic features and formations. For instance, there is the vast Tibetan Plateau which consists of sedimentary deposits, which are thousands of feet thick, located presently at an elevation of three miles above sea level." He goes on to say, "Creationists attribute modern-day topography to sudden and supernatural causation after the Flood (Psalm 104:6-9) which helps to explain how sedimentary strata (formed during the Flood) have come to be lifted thousands of feet above sea level in the mountainous regions of the earth."

Furthermore, creationists believe that only a hydraulic catastrophe of global scope could have created the world's deposits of sedimentary rock. The abundant amounts of rocks of considerable size in the various strata on a worldwide basis, indicates a period of very forceful hydraulic activity. Such a period of intense hydraulic activity is consistent with the conditions that would have existed during the great Flood of Genesis. The extensive movement of gravels, conglomerates, and individual rocks the size of boulders is not consistent with the low intensity, hydraulic activity that is said to have characterized earth history, according to uniformitarianism. This is one of the reasons that creationists reject the uniformitarian interpretation.[16]

Fossilization, Evolutionary Progression and Fossils of Mammals

This section will deal with the fossils, themselves. According to Dr. Gish, "for a plant or an animal to become a fossil (except under very special circumstances), it must be buried almost immediately after it dies. If an animal dies, and then just lies around on the ground or floats around in the water, it never becomes a fossil." If dead plants or animals are exposed a number of things can happen. They can decompose due to the action of bacteria or due to oxidation. They could be eaten by worms, insects, or various scavengers. These dead plants and animals might also be dissolved by the action of chemicals in soil and water.[17]

The process of fossilization involves the replacement of skeletal remains or perhaps plant material by minerals in the groundwater. For example, when a tooth from a dead animal comes in contact with this groundwater, the mineral that is dissolved in the water gradually replaces the original material of the tooth. That tooth will eventually become as hard as a rock when replacement is complete. Minerals such as silica, calcite, and pyrite are involved in this process. Fossilization has been observed to occur within a matter of decades. Fred J. Meldau wrote, in *Why We Believe In Creation Not In Evolution*, "people think that it takes many thousands of years to produce a fossil. This is not necessarily so. There are fossil men in the ruins of Pompeii overwhelmed by an eruption of Mt. Vesuvius, in 70 A.D." Meldau went on to say, "A fossilized Mexican sombrero was found not many years ago; it couldn't have been over 200 years old!"[18]

According to Dr. Velikovsky, "the explanation of the origin of fossils by the theory of uniformity and evolution contradicts the fundamental principle of these theories: Nothing took place in the past that does not take place in the present. Today no fossils are formed." He also wrote, "Millions of buffaloes have died natural deaths on the prairies of the West in the more than four hundred years since the discovery of America; their flesh has been eaten by scavengers or putrefied and disintegrated; their bones and teeth resisted for a while the decaying process, but finally weathered and crumbled to powder. No bones of these dead buffaloes became fossils in sedimentary rocks and scarcely any are found in a state of preservation."

As mentioned already, rapid burial is almost always required for an animal to become a fossil. This fact presents a big problem as far as the fossilization of fish is concerned. According to Velikovsky, "when a fish dies its body floats on the surface or sinks to the bottom and is devoured rather quickly, actually in a matter of hours, by other fish. However, the fossil fish found in sedimentary rock is very often preserved with all its bones intact. Entire shoals of fish over large areas, numbering billions of specimens, are found in a state of agony, but with no mark of a scavenger's attack."[19]

Dr. Velikovsky's very insightful remarks show that the uniformitarian interpretation is inadequate to account for the billions upon billions of fossils that are buried within the sedimentary rocks. This is particularly true of the fossilization of fish, which is virtually impossible under normal conditions. The mass burial of large shoals (schools) of fish is really only explainable in terms of the Biblical Flood. The Flood would have created conditions in which enormous quantities of sediments would have poured into antediluvian lakes and seas. Vast numbers of fish and other aquatic organisms would have been caught in these suffocating streams of sediments and quickly buried.[20]

Now, if fossils are very, very seldom, if ever, formed under normal conditions, then the billions of fossilized organisms in the sedimentary rocks must be the result of highly abnormal conditions. Or in other words, all of those fossils reflect the occurrence of catastrophic conditions at some point in the Earth's history. Furthermore, since very large numbers of fossils are observed all over the world, these catastrophic conditions must have been global in extent. There is no dispute concerning the fact that the world's sedimentary rocks were primarily formed from sediments that were carried by water. There is also no disagreement over the fact that rapid burial, by sediments, is necessary for fossilization. So at some point in the past, there must have been a global hydrologic catastrophe. This catastrophe must have been of sufficient intensity that millions of tons of sediment were carried by moving waters. These sediments were necessary for the rapid burial of the billions of organisms that were fossilized. Creationists believe that this catastrophe, in which the billions of animals and plants were covered by sediments, was the great Flood of Genesis.

Dr. Morris wrote, "With the primary purpose of the Deluge being to destroy all life on the earth (at least on the dry land) except the Ark's passengers, there must have been uncounted multitudes of living creatures, as well as plants, trapped and eventually buried in the moving masses of sediments, and of course under conditions eminently conducive to fossilization. Never before or since could there have been such favorable conditions for the formation of fossiliferous strata." The rapid burial, by sediments, is the first step of fossilization.[21]

Generally speaking, there is a progression of organisms, from simple to more complex, in the fossil record. The lower levels of rock strata, for the most part, contain relatively simple organisms. Then, as you observe higher levels of sedimentary rock you will usually see more complex organisms. This could be called a general rule but many, many exceptions have been found. In the early days of paleontology this general progression, towards more complex organisms as you go up in the geologic column, was mistakenly interpreted as being evidence of evolution. However, since the mid-nineteenth century so many exceptions to this general rule have been observed that such an interpretation is no longer tenable. Creationists believe that this tendency, towards more complexity as you go up in the geologic column, is due to the "hydraulic sorting action" of the great Flood.[22] The following passage, from *The Genesis Flood*, is an explanation of this sorting action.

Dr. Morris, who earned a Ph.D. in engineering from the University of Minnesota, wrote, "Particles which are in motion will tend to settle out in proportion mainly to their specific gravity (density) and sphericity. It is significant that the organisms found in the lowest strata, such as the trilobites, brachiopods, etc., are very 'streamlined' and are quite dense. The shells of these and most other marine organisms are largely composed of calcium carbonate, calcium phosphate, and similar minerals, which are quite heavy-heavier, for example, than quartz, the most common constituent of ordinary sands and gravels. These factors alone would exert a highly selective sorting action, not only tending to deposit the simpler (i.e. more nearly spherical and undifferentiated) organisms nearer the bottom of the sediments but also tending to segregate particles of similar sizes and shapes, forming distinct faunal stratigraphic 'horizons' with the complexity of structure of the deposited organisms, even of similar kinds, increasing with increasing elevation in the sediments."

There are other factors that have contributed to this superficial resemblance to an evolutionary progression. For one thing, in a global hydrologic/hydraulic catastrophe it would be expected that the first creatures to be covered by sediments would be marine organisms. The sea bottoms of the antediluvian seas would have been the first areas affected when the Flood of Genesis began. As a matter of fact, rock strata with fossils of marine creatures are usually found to be the lowest strata in a real-world geologic column. For example, Cambrian rocks are ostensibly the most ancient fossiliferous strata. Of the 1500 species of invertebrates that have been found in Cambrian rocks, all are marine invertebrates. About 60% of Cambrian fossils are trilobites.

Furthermore, it seems sensible and predictable that vertebrates would generally be found higher in the column because of their greater mobility. Generally, land plants and animals would be expected to be overcome by rising waters and buried by sediments some time after the trilobites. The trilobites had a shallow marine habitat. The exact location of any type of organisms, in the rock strata, would depend, largely, on the elevation of their original habitat. That fossilized amphibians would appear lower in the rock strata than fossilized mammals is predictable on the basis of habitat. That fossilized reptiles would normally appear lower in the rock strata than mammals is predictable on the basis of differences in mobility. This general rule of progression, that is reflected by the fossil record, actually corroborates the Flood account.[23]

Now, we'll make a sweep up through the geologic column and look at some problems with the uniformitarian/evolutionary interpretation. One of these problems is the gaping chasm in the fossil record that occurs between pre-Cambrian and Cambrian rocks. The great majority of Precambrian rocks only contain evidence of microscopic, single-celled organisms. But just above the Precambrian rocks, in Cambrian rock strata, there is abundant evidence of complex multicellular invertebrates. From an evolutionary standpoint, this is a big jump for which there are no transitional forms. Fred J. Meldau wrote, "All geologists note this strange phenomenon: THERE IS LITTLE FOSSIL RECORD BEFORE THE CAMBRIAN PERIOD (first of the Paleozoic era) - long after the fossil record SHOULD have appeared if evolution is correct!" Mr. Meldau goes on to say, "The inference from the fact that there are almost no pre-Cambrian fossils is, life began on earth suddenly and in great variety."[24]

The sudden and profuse appearance of fossilized organisms in Cambrian rocks is often referred to as the Cambrian explosion. The evolutionists claim that this explosion was the result of a period of tremendously accelerated evolution. This period is said to have begun about 570 million years ago and is supposed to have had a duration of at least 10 million years. The accelerated evolution concept is an attempt to explain why you see very little evidence of life in Precambrian rock and then see such an impressive array of complex invertebrates (e.g. clams, brachiopods, sponges, and trilobites) in Cambrian rock strata. Dr. Gish wrote, "Many billions times billions of the intermediates would have lived and died during the vast stretch of time required for the evolution of such a diversity of complex organisms. The world's museums should be bursting at the seams with enormous collections of the fossils of transitional forms. As a matter of fact, not a single such fossil has ever been found!"

The evolutionists claim that evolution proceeded so rapidly, at that stage of geologic history, that there wasn't sufficient time for the transitional forms to have left a detectable fossil record. But wait a minute! The observable fossil record was supposed to have been the best evidence of evolution in the first place. But in regards to the Cambrian explosion, they seem to be saying that the absence of a fossil record is the evidence of accelerated evolution. Modern genetics does not support this concept of accelerated evolution. In fact, evolutionists have always said that no one has ever witnessed one species evolving into another species because evolution works so slowly. Furthermore, the genetic apparatus of a certain kind of animal is completely devoted to reproducing that kind although there can be wide variation within that particular kind.[25]

The concept that evolution occurs in bursts, or jumps, is referred to as punctuated equilibrium This is a fairly drastic departure from traditional Darwinism. Darwin, himself, actually advocated a very gradual version of evolution. Charles Darwin was aware of the paucity of fossils that might be considered to be intermediates. The evolutionist Gould has basically admitted that paleontologists have been rather secretive about the lack of transitional forms. Darwin believed that the intermediates would eventually be found in the fossil record. However, it has been 140 years since the publication of *The Origin of Species* and those intermediates still haven't been found.[26]

Now, before we go higher in the geologic column it should be mentioned that there is a group of multicellular organisms that are said to be Precambrian multicellular animals called metazoans. The Precambrian metazoans are called Ediacaran Fauna. These fossils, of soft-bodied creatures, were originally discovered near Adelaide, Australia. The fossilized remains of these Precambrian metazoans were found in rocks that were said to be around 650 million years old. Since that initial discovery, Ediacaran Fauna have been found in other parts of the world.

According to Gish, "these discoveries do not alleviate the problem for evolution theory. These creatures are in no way intermediate between single-celled organisms and the complex invertebrates previously found in Cambrian rocks. They *are* complex invertebrates. Furthermore, it has been recently established that the creatures of the Ediacaran Fauna are not the same as the worms, coelenterates, and echinoderms of the Cambrian. In fact, they are so basically different that it has been stated unequivocally that they could not possibly have been ancestral to any of the Cambrian animals."[27] Very prominent evolutionists have plainly said that there is no evolutionary connection between the Ediacaran Fauna and the invertebrates of Cambrian rock strata. Seilacher and Gould have shown that the Ediacarans were not, at all, related to the shelled animals of Cambrian rocks. This is because the Ediacarans are so different from what came later.[28]

Evolutionists have tried to explain away the problem of lack of transitional forms in Precambrian rocks by saying that a fossil record wasn't left because Precambrian organisms did not have hard parts. But Morris and Morris wrote, "This is a poor excuse. None of the Precambrian fossils had hard parts. Furthermore, many of the larger Cambrian fossils did not have hard parts (e.g. jellyfish), yet their fossils are available in considerable numbers."[29] According to Johnson, "the Ediacarans actually demolish a standard Darwinist explanation for the absence of pre-Cambrian ancestors: that soft-bodied creatures would not fossilize. In fact many ancient soft-bodied fossils exist in the Burgess Shale and elsewhere."[30] The important thing to remember is that the Ediacaran Fauna do not even begin to bridge the enormous gap between the microscopic fossils of Precambrian rock strata and the fossilized invertebrates of Cambrian rocks.

There is another huge gap in the fossil record that exists between the invertebrates and fish. This gap really proves, beyond any doubt, that evolution did not happen. This gap is between the fossils of the invertebrates and the vertebrate fossils. The evolutionists claim that fish were the first vertebrates. They claim that the evolutionary development of fish, from invertebrates, took millions upon millions of years. Obviously, there are drastic differences in body structure between the Cambrian invertebrates and fish. If such evolution did occur it would have involved multitudes of transitional forms. The existence of these transitional creatures should be well documented (so to speak) in the fossil record. However, such fossilized intermediates cannot be found. The evolutionary links between invertebrates and fish are missing.[31]

According to evolutionary doctrine, there was a chordate stage between invertebrates and vertebrates. Dr. Gish wrote in chapter four of *Evolution: The Fossils Still Say No!*, "The transition from invertebrate to vertebrate supposedly passed through a simple chordate state, that is, a creature possessing a rod-like notochord. Does the fossil record provide evidence for such a transition? Not at all." He went on to say that "some evolutionists boastfully cite a fossil chordate, Pikaia, as an intermediate. One single fossil chordate as their 'evidence' for the evolution of invertebrate into vertebrate! But if evolution is true, millions of undoubted intermediates showing the gradual evolution of fishes from its invertebrate ancestor should crowd museum shelves and be on display for any doubters to see."

Besides the fact that evolutionists can't account for the gap between invertebrates and fish, there are also gaps between the different orders of fish. They make a sudden appearance in the fossil record. The fossil record does not reflect that they have evolved.[32] Evolution is dead without the necessary transitional forms to show evolutionary pathways.

The evolutionists claim that fish evolved into amphibians. But they don't have the necessary fossilized intermediates. The evolutionary transformation of fish to amphibians is said to have taken about 30 million years, but not one "fishibian" has been found in the fossil record. It was once thought that a fish called coelacanth was an intermediate. This fish is found in Cretaceous strata, but was discovered to be still living in 1938.[33] It was also suggested that amphibians evolved from the lungfish. Another fish, which is found in Carboniferous strata, called Rhipidistian is said to have been an intermediate between fish and amphibians. There doesn't seem to be a consensus among paleontologists concerning any of these three alleged ancestors of amphibians. This confusion and difficulty, as far as amphibian evolution is concerned, is a consequence of the lack of transitional forms in the fossil record.

Dr. Gish makes an important observation concerning the supposed evolution of amphibians when he asked "how could a fish or the alleged proto-amphibian survive on land before his many sensory structures had undergone extensive reorganization in order to adapt to the physical and chemical differences between water and air? Keep in mind that each sense organ had to function correctly from the very first, each change had to occur in the right sequence and be coordinated with all others, and all of this had to take place via random or accidental changes or mutations of the genes."

Many other important physiological changes would have had to take place to enable the nascent amphibian, or part-way amphibian, to survive on land.[34] Furthermore, there is no evidence in the fossil record of any intermediates between the different orders of amphibians. All of the orders of amphibians are very different from each other with no evidence of any evolutionary connection. The same thing could be said for all the orders of fishes as well as all the orders of reptiles. These facts are not seriously disputed by vertebrate paleontologists. The actual fossil record just does not have the transitional forms that are necessary to prove the validity of evolution. According to Morris and Morris "the transitional forms so essential to evolutionary theory are missing. These facts, on the other hand, fit the creation model perfectly!"[35]

Amphibians are supposed to have evolved into reptiles. Yet there are profound differences between these two classes of animals. Amphibians lay their eggs in water. When the young of a frog hatch they must live in the water as tadpoles, breathing through gills. Later on in a frog's life, it develops lungs and can move about on land, breathing air. By contrast, reptiles lay their eggs on land. Reptilian eggs are much different than amphibian eggs. Baby reptiles never have gills and can breathe air at birth.

Amphibians must stay fairly close to water while some reptiles survive well in deserts.

The development of the reptile egg presents a real problem for evolutionists. The egg of a reptile is far more complex than that of an amphibian. For one thing, reptile eggs contain an amnion, which is a tough membrane that forms a sac. This sac contains a watery fluid in which the embryo floats. The reptile egg has a reservoir for waste products which the amphibian egg doesn't have. The reptilian egg has its food supply contained in a yolk sac, which the amphibian egg does not have. The reptile egg is surrounded by a shell for protection. Yet this shell is porous enough to allow an exchange of gases with the outside air.

The evolutionists are faced with the formidable task of explaining how the very complex reptile egg evolved, from the amphibian egg, by a series of random, accidental, genetic changes. These changes had to be concurrent with certain required alterations in the reproductive organs of the proto-reptile. These other reproductive changes would have also been due to random and accidental genetic changes. The evolutionists must also explain how all the transitional stages were completely functional. Each successive intermediate, or transitional stage must also have had a competitive advantage over the previous intermediates. The evolutionists must explain all these changes. For example, it is difficult to imagine how the amnion came to be since there is nothing, at all, similar to it in the amphibian egg.

Some evolutionists believe that a certain creature, called Seymouria, was transitional between reptiles and amphibians. However, there are two very big problems with calling Seymouria a transitional form. For one thing, it appears in strata that is supposed to be younger than the earliest reptiles appear in. According to the standard geologic column/time scale, Seymouria came along about 25 million years too late to be an ancestor to reptiles. Another problem with calling Seymouria an intermediate between amphibians and reptiles is that it had a definitely amphibian reproductive system.[36]

Concerning the supposed amphibian to reptile transition, Johnson wrote, "The most important difference between amphibians and reptiles involves the unfossilized soft parts of their reproductive systems. Amphibians lay their eggs in water and the larvae undergo a complex metamorphosis before reaching the adult stage. Reptiles lay a hard shell-cased egg and the young are perfect replicas of adults on first emerging. No explanation exists for how an amphibian could have developed a reptilian mode of reproduction by Darwinian descent."[37] The fossil record simply does not reflect that amphibians are ancestral to reptiles.

One of the most interesting kinds of reptiles are the turtles. These cold-blooded vertebrates have a bony, protective shell, which is also called a carapace. Only turtles have this type of protective covering. Turtles are also unique in other respects. According to Gish, "the positions of the ribs and the pectoral and pelvic girdles are completely reversed, with the ribs outside, or dorsal, to the girdles, which lie inside, or ventral, to the ribs. Here again, because of the amazingly unique structure of turtles, if they evolved from ordinary reptiles of some kind, it should be a rather easy task to find the transitional forms that would trace the evolutionary pathway from ancestral reptile to turtles." According to evolutionary theory, turtles and other groups of reptiles evolved from a common reptile ancestor. However, no fossilized intermediates have been found that connect turtles and other reptiles to any reptile ancestor. In other words, the fossil record basically reflects that turtles have always been turtles.

Reptiles of the order Squamata include lizards and snakes. According to evolutionary doctrine, the snakes gradually evolved from lizards. However, there are differences between the members of these two suborders. For one thing, there is the peculiar elongation of the snakes trunk. Some snakes have hundreds of vertebrae. These vertebrae seem to have been altered to permit the snake's slithering movement. Even though they supposedly developed from lizards, snakes do not exhibit any trace of forelimbs or pectoral girdles. Some snakes do have hind limbs and a small pelvic girdle. The hind limbs and pelvic girdles have been called vestigial structures by the evolutionists. However, these structures are there for good reasons. Specific muscles are anchored by the snake's pelvic bones. The hind limbs can be used to enhance movement and also serve a purpose during mating.

Dr. Gish wrote, "One would expect to find a series of transitional forms between lizards and snakes documenting the gradual multiplication of vertebrae, loss of limbs, modification of the skull, and other significant changes as a lizard was transformed into a snake. Also, the question must be answered, why would an enormously successful lizard, via any selection process, undergo gradual changes resulting in losing its limbs, exchanging its efficient mode of locomotion for stages intermediate between lizards and snakes that certainly would be very inefficient?" Paleontologists have referred to the fossil record, concerning snake evolution, as being "very fragmentary." This is really a kind of euphemism to conceal the fact that there are no transitional forms to establish an evolutionary connection between snakes and lizards.[38]

Now, let's look at another group of reptiles that are now extinct. According to Gish, "there were other kinds of rather frightening creatures that lived at the same time with the dinosaurs. Apparently, they all died out some time after the Flood, just as the dinosaurs did. These were the flying reptiles." At one time, scientists didn't think that this group of creatures actually flew. It was believed that they just glided. It was thought that these flying reptiles had to climb up to high places in mountainous areas. Then they had to jump off at just the right time to catch a warm updraft of air. However, scientists now think that these reptiles could actively fly as birds do.

The flying reptiles included the little *Rhamphorhyncus* which was about 18 inches long. This creature, like all the flying reptiles had a narrow, extended skull. *Rhamphorhyncus* is believed to have had a diet of fish, which it speared with teeth that pointed forward. Another flying reptile, called *Ptreranodon*, had a wingspan that averaged 23 feet but its trunk was about the size of that of a turkey. *Quetzalcoatlus* was a huge flying reptile that was named after one of the gods of Aztec mythology. It had a wingspan of about 50 feet, which exceeds that of the military aircraft called F-4. It is believed to have caught fish that were just beneath the surface of the ocean. All of the flying reptiles had hollow bones that were very light. They had leathery wings that resembled those of a bat. The flying reptiles had hands with an extremely long fourth finger. That long fourth finger was the forward edge of the wing.[39]

The evolutionists would have us believe that these flying reptiles evolved from land reptiles. Theoretically, these creatures gradually developed wings. This means that at some point the flying reptiles supposedly had wings that were about 25% of their eventual size. But imagine how cumbersome such wings would have been. Those wings couldn't have allowed flight due to their small size. Reptiles would have had to drag those appendages around. These one quarter sized wings would have prevented them from either catching prey or escaping danger. Such an evolutionary experiment would have completely failed. It almost goes without saying that there are no fossils anywhere of reptiles that had wings at any stage of transition. The flying reptiles did not evolve at all. These amazing creatures were created by God.[40]

Evolutionists claim that birds evolved from reptiles. The skeletons of birds and reptiles are said to be similar. Once again, transitional forms are necessary to establish the evolutionary pathway between reptiles and birds. Dr. Huse, who has earned both the Ph.D. and the Th.D. degrees, wrote, "The only fossil ever found and proposed to be a transitional link between these two classes is the famous *Archaeopteryx*. It is difficult to understand (if you believe in evolution) why only one fossil has ever been uncovered that might be transitional between these two groups of different animals especially since evolutionists estimate an eighty million year developmental time span from reptiles to birds."[41]

The *Archaeopteryx* was found in sedimentary strata that was designated as Jurassic. Evolutionists claim that this creature resembled a reptile for a number of reasons. These reasons include the following; its tail is long and bony, it has claws that are on its wings, its bill has teeth, it has an unfused backbone, and that most *Archaeopteryx* fossils don't seem to have a breastbone with a keel (ridge to which muscles are attached). But Dr. Parker wrote, "The reptile-like features are not really so reptile-like as you might suppose. The familiar ostrich, for example, has claws on its wings that are even more 'reptile-like' than those of *Archaeopteryx*. Several birds, such as the hoatzin, don't have much of a keel. No living birds have socketed teeth, but some fossil birds did. Besides, some reptiles have teeth and some don't, so presence or absence of teeth is not particularly important in distinguishing the two groups."[42]

Furthermore, the evolutionists have a problem explaining why the *Archaeopteryx* had feathers that are completely formed. According to evolutionary doctrine, feathers began as the frayed scales of reptiles. But an examination of *Archaeopteryx* fossils reveals that its feathers do not, at all, resemble frayed scales. A truly intermediate stage would be expected to have unusual feathers that somehow resembled scales. Its feathers should not be completely bird-like. Yet the *Archaeopteryx* has feathers that are fully like that of a bird. In fact, a halfway stage between the reptilian scale and the feather has not been found anywhere in the fossil record. No kind of transitional scale-feather is known to have ever existed. The evidence that the *Archaeopteryx* represents an evolutionary link between reptiles and birds is very weak, at best.[43]

Because of the fact that *Archaeopteryx* is the leading candidate as a transitional form, the supposed evolution of reptiles to birds isn't even close to being proven. Fred J. Meldau wrote, in that creationist classic *Why We Believe In Creation Not In Evolution*, that "there is ABSOLUTELY NO EVIDENCE WHATEVER OF THE SUPPOSED 'GRADUAL CHANGE' OF SCALES INTO FEATHERS, or of the development of wings, the loss of teeth, the development of exceptional sight and the hundred and more other colossal differences between birds and reptiles. A partly developed organism (such as a bird's wing, claw, bill, feather, etc.) IS OF NO VALUE WHATEVER TO A LIVING ANIMAL, and such 'partly developed' organisms are nowhere found in nature. Evolution exists ONLY in the minds of its devotees." Meldau goes on to say, "Slight, gradual, random mutations do not account for such drastic changes involved in, 'the acquisition of flight in birds', for, to be successful, the entire body had to be rebuilt at the same time in order to make flight possible! The phenomenon of radical changes such as the development of flight in birds, precludes the idea of gradual changes by random mutations. The only way a bird could possibly come to being is by a SUDDEN CREATION."[44]

The mammals are also said to have evolved from reptiles. But are there recognizable transitional forms in the fossil record to establish an evolutionary pathway between reptiles and mammals? Evolutionists point to certain reptiles with characteristics that were similar to mammals. All of the various mammal-like reptiles were very numerous in the past but are now extinct. Because the mammal-like reptiles aren't living now their soft parts can't be examined. It is this soft biology by which living reptiles and mammals are now differentiated. The different orders of mammal-like reptiles seem to have thrived in their particular environments. These orders included the therapsids, theriodonts, cynodonts, and pelycosaurs, among others.

Based on the assumption of the validity of evolution, it seems reasonable that the reptiles with some mammalian characteristics must have been ancestral to mammals. However, creationists do not accept this assumption. Morris and Morris wrote, "There is no reason whatever to assume that the mammal-like reptiles were animals in the process of being transformed from reptiles to mammals. They originated simultaneously with the other orders of reptiles, with no clear indication of ancestry."[45]

This is another case where the transitional forms just aren't there. Dr. Gish wrote, "In their attempts to establish an evolutionary tree or phylogeny for the mammal-like reptiles and the mammals, evolutionists rely almost entirely on similarities to link these creatures in an evolutionary scenario. They are forced to do this because of the lack of transitional forms required for their hypothetical evolutionary ladder." The fact that one kind of animal has some characteristics in common with another kind of animal doesn't necessarily indicate an evolutionary link.[46] That different kinds of animals have characteristics in common actually reflects an overall design in nature. One might say that this is evidence of a master plan. The study of design in nature is called teleology.

Creationists believe that God designed certain animals to have a "mosaic", or mixture, of traits. One such animal is the duck-billed platypus. According to Huse, "the platypus is an Australian mammal which has fur and nourishes its young with milk, like mammals. Their young hatch from eggs like reptiles, and they also have webbed feet and a flat bill like a duck. It has pockets in its jaws to carry food and a spur on its rear legs which like a snake's fang, is poisonous. And amazingly enough, the platypus even uses echo location like dolphins!" Evolutionists claim that the platypus is an intermediate between birds and mammals. Yet such a bizarre combination of characteristics seems to eliminate that possibility.[47]

As with the platypus, the reptiles that were mammal-like possessed a mosaic of characteristics. For example, the theriodont fossils show that they were animals that had some features that were clearly mammalian and some features that were definitely reptilian. The point is that theriodonts didn't have particular traits that were half-mammalian and half reptilian. A true transitional form would have traits that reflected a progression, such as the frayed scale feathers of a supposed reptile-bird intermediate. Dr. Gish wrote, "In a great number of cases we should be able to trace, via transitional forms, the origin of each distinct kind, as the many different types of mammal-like reptiles evolved, until we arrive at the final stage, a creature no longer merely mammal like but 100% mammal." Gish goes on to say that the "fossil record produces neither the evidence of gradual change nor the transitional forms predicted on the basis of evolution."

One evolutionist, Kemp, has admitted that there are numerous gaps in the fossil record of the mammal-like reptiles. Dr. Gish wrote, "Kemp asserts that the reptile to mammal transition is the best documented case for evolution, but then must admit that hypothetical transitional forms must be constructed because intermediate forms between known groups are almost invariably unknown! The first mammal-like reptiles appear in the rocks of the Upper Pennsylvanian, allegedly about 350 million years ago, and supposedly became extinct at the end of the Triassic. Thus, evolutionists believe the mammal-like reptiles spent almost 200 million years evolving before reaching mammalian status." Countless numbers of transitional forms should be in the fossil record to trace the evolutionary pathway between reptiles and mammals. Yet such fossilized intermediates are non existent.

T. S. Kemp claims that the transitional forms are missing due to rapid bursts of evolution at the species level. Thus, Kemp accepts the punctuated equilibrium model of evolution. It must be recognized that when they attempt to explain the lack of transitional forms by these hypothetical bursts of evolution, the evolutionists are really absurdly presenting the lack of evidence as evidence. When one asks them why there are no fossilized intermediates, one is told that it is because of these bursts of evolution. Yet the only evidence that such bursts of evolution ever took place is that there are no fossilized intermediates. So the punctuated equilibrium model is really just an illusion.

There are profound differences between reptiles and mammals. For example, reptiles are far different from mammals as far as temperature control is concerned. According to Gish, "the physiology of mammals must permit the maintenance of a constant, relatively high body temperature This characteristic, called endothermy, is accomplished by heat production through a cellular metabolic rate that is about seven times that of a cold-blooded (ectothermic) reptile. Endothermy requires a finely adjusted highly complex biological organization that is found in mammals but not in reptiles."

Homeostasis is a term that refers to a mammal's capacity to regulate its internal environment in order to endure environmental vicissitudes. Temperature control is just one aspect of homeostasis. Water retention is another aspect of homeostasis. In fact, there are 27 processes and anatomical structures that must be perfectly coordinated to maintain the internal environment. According to evolutionary doctrine, all of these structures and processes are supposed to have evolved in a synchronous fashion as a result of random genetic accidents. Furthermore, this is said to be true for the mammalian organism as a whole.

All aspects of a mammal's physiology and anatomy are highly integrated and amazingly complex. This is particularly true of the 27 processes and structures that are required for homeostatic control. This means that the anatomy and physiology of the mammal must have evolved very gradually (assuming that the mammals evolved). The mammal's highly integrated anatomy and physiology would have evolved in a completely coordinated way. No process or structure could have evolved separately from all the others. So besides the homeostatic adjustment systems, all the various systems and structures of the entire mammal must have evolved in a perfectly synchronized manner, via random genetic accidents, if evolution is valid.[48] In fact, according to evolutionary doctrine, all of the systems and structures of every organism that ever lived evolved in a coordinated way, via random genetic accidents. It takes faith to believe in evolution.

Fossils of mammals were found in Cretaceous rocks in the Gobi Desert. The fossils of 187 mammals were found by a group of American and Russian scientists. Dinosaur fossils are also found in Cretaceous rocks. Concerning these fossils, Dr. Gish wrote that 'this discovery may force evolutionists to rethink the way they portray mammals, since this find reveals that mammals were much more extensive and diverse before dinosaurs became extinct and thus, Novacek says, were not direct competitors of dinosaurs. There have been found in the Gobi Desert area fossils of dinosaurs, mammals, lizards, crocodiles, and turtles, so the emerging picture of ancient life is looking more and more as visualized by creationists."[49]

For a long time the evolutionists claimed that the fossilized horse remains provided conclusive proof that evolution really occurred. The fossil horse series has been one of the most well known examples of "proof" of evolution. However, Dr. Huse wrote, "Far from being the well established fact, which evolutionists portray it to be, the horse series is plagued with many serious problems and is actually nothing more than a deceitful delusion."

The fossil horse series has a number of discrepancies. One of these discrepancies is that there are no fossilized intermediates to establish an evolutionary link between the horses of the series. For example, *Eohippus* is said to be the first horse of the series. However, *Eohippus* is not connected by any fossilized intermediates to its reputed ancestors, the *condylarths*. *Condylarthra* was an order of hoofed mammals that is said to have been prominent in the Paleocene epoch. The condylarths are now extinct. Another problem is that the fossil record does not really reflect an evolutionary progression from many small toes to just one large toe. Furthermore, Dr. Huse wrote, "A complete series of horse fossils in the correct evolutionary order does not exist anywhere in the world." He goes on to say, "The fossil horse series starts in North America, jumps to Europe, and then back again to America."[50]

The evolutionists claim that *Eohippus* evolved to become the modern horse, *Equus*. Other horses (e.g. *Mesohippus*, *Miohippus*, and *Pliohippus*) are said to have been intermediate stages of horse development. Supposedly, as this succession of horses increased in size, the number of toes was devolving from five to just one. As previously mentioned, the fossil horse series has been considered to be a decisive proof that evolution occurred. However, problems with the fossil horse series have caused many scientists to question its validity.

Besides the discrepancies already mentioned, there is a big problem with *Eohippus*. This alleged ancestor of the modern horse actually resembled a hyrax more closely than it did a horse. The hyrax is a somewhat rabbit-like hoofed mammal in the order *Hyracoidea*. The coney of Proverbs 30:26 may be a hyrax. So *Eohippus* doesn't really belong in a horse series. In fact, *Eohippus* doesn't differ very much from a hyrax. The evolutionists don't like it when the fossilized remains of creatures that are found in rock strata, which the evolutionists say is many millions of years old, are found to closely resemble living animals.

Research in paleontology has thrown doubt on the other members of the horse series. For example, the three toed *Merychippus* has been found in the same rock strata as *Pliohippus* in Oregon. The *Pliohippus* has just one toe and has also been found in exactly the same sedimentary rocks as the *Hipparion*. *Hipparion* has three toes and this discovery was made in the Great Basin area. It is certainly fair to say that the *Equus* family tree has turned into something of a bush. All members of the fossil horse series seem to have lived at about the same time. This has been admitted by the paleontologist Stanley.[51] According to Whitcomb and Morris, the fossil horse series was "constructed mainly on the basis of evolutionary presuppositions as to the possible relationships between these various creatures. The series thus constructed is thereupon submitted as proof positive for the evolution of the modern horse!"[52]

The marine mammals have presented a frustrating problem to the evolutionists. Mammals are said to have evolved from reptiles and reptiles live on land. So the evolutionists have really had to put their thinking caps on to get those marine mammals back in the water. Consider the whale. The evolutionists claim that about 60 million years ago some hairy mammals went into the water. These mammals may have been seeking sanctuary or were perhaps looking for food. For some unknown reason these mammals decided to stay in, or very near, the water. Supposedly, evolutionary changes occurred over a period of millions of years. The front legs are said to have become flippers, but the back legs vanished. The hair is said to have somehow become a layer of blubber. The nostrils are imagined to have traveled from the front to the top of the head. The body and tail grew to enormous size, supposedly. Dolphin evolution is said to have taken place in a similar way.[53]

According to Gish, "Wursig has suggested that dolphins may have evolved from land mammals resembling the even-toed ungulates of today such as cattle, pigs, or buffaloes. It is quite entertaining, starting with cows, pigs, or buffaloes, to attempt to visualize what the intermediates may have looked like. Starting with a cow, one could even imagine one line of descent which prematurely became extinct, due to what might be called an 'udder failure'." Dr. Gish goes on to say, "The consensus of evolutionists today has settled on the hairy, four-legged carnivorous mammal, *Mesonyx*, as the probable ancestor of all marine mammals. This opinion is based largely on dental evidence and shape of the skull. The mesonychids were wolf-like, hoofed carnivores that, as far as anyone knows, never went near water except to drink." So the evolutionists would have us believe that once upon a time, long ago, the whales looked like wolves.[54] These are probably the same wolves that lived in the enchanted forest, with the elves. Yet the most amazing thing is that evolutionists actually have the nerve to call this wolf-to-whale scenario science.

In conclusion, the fossil record does not reflect that mammals or, for that matter, any of the other classes of animals have evolved. The transitional forms, that are necessary to establish the evolutionary pathways, are not in the fossil record. Evolution is dead without these transitional forms. That the layers of sedimentary rock strata were deposited over a period of hundreds of millions of years was an assumption made by scientists in order to accommodate evolutionary theory. By contrast, creationists believe that the world's fossiliferous strata exist as the result of a relatively short period of deposition during a global hydrologic/hydraulic catastrophe. Creationists believe that this global catastrophe was the great Flood that is described in the seventh and eighth chapters of Genesis.

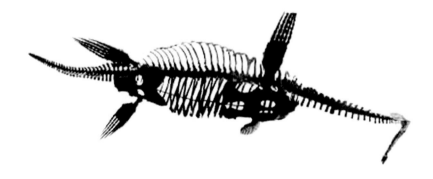

ELASMOSAURS Platyurus

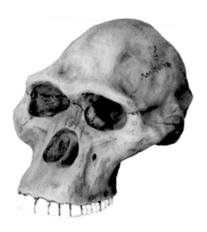

Australopithecus africanus

Homo Erectus

Neanderthal Man

CHAPTER 8

HUMAN FOSSILS

Australopithecus afarensis is said to be mankind's earliest known primate ancestor. Most of the current speculation about the origin of humans concerns the fossilized remains of the australopithecines, which are now extinct. The best known australopithecine specimen is called Lucy, which is a fossilized skeleton that is about 40% complete. D.C. Johanson discovered these remains in Ethiopia. Johanson was in the Afar area of that country from 1972 to 1977.[1] It was surprising to paleonanthropologists that so much of this skeleton was recovered in view of its alleged age of over three million years. At such an age, only a few fragments are expected. According to Dr. Lubenow, "from their evaluation of Lucy and fossils like her, Don Johanson and Tim White (University of California, Berkeley) decided in 1979 that Lucy was our oldest known direct ancestor."[2]

Dr. Gish wrote, "Johanson discovered the knee joint of a small primate which he at first assumed was that of a monkey. After fitting the parts together and noting the angle the joint appeared to form, he declared that it was the knee joint of a hominid, that is, of a creature intermediate between ape and man. He furthermore believed, on the basis of the fossils of animals that had been found in the area, that his fossil knee joint was three million years old. He thereupon declared on the spot that he had discovered a three million year old human ancestor." This discovery was in the fall of 1973, but these remains were not those of the famous Lucy. The fossilized skeleton, that was called Lucy, was discovered in the fall of 1974. Lucy was about 3 1/2 feet tall, with a cranial capacity of about 400 c.c.[3] Dr. Lubenow described this creature as being "chimp like."[4]

One of the reasons that Lucy was said to be in the ancestry of mankind concerned the issue of bipedalism. This term refers to the ability to walk upright in the manner of humans. Johanson claimed that Lucy walked upright. However, according to Huse, "the chimpanzee spends a considerable amount of time walking upright. Thus, there is no valid scientific basis for a conclusion of bipedalism in Lucy."[5] Gibbons can also walk upright, while walking along large branches of trees.[6]

Australopithecus afarensis is said to have evolved into *Australopithecus africanus*. Dr. Gish wrote that "the first find of this creature was by Raymond Dart in 1924 to which he gave the name *Australopithecus africanus*. He pointed out the many ape-like features of the skull, but he believed that some features of the skull, and particularly of the teeth, were man-like. The name *Australopithecus* means 'southern ape', but after Dart examined the teeth further, he decided *A. africanus* was a hominid."[7]

Raymond Dart was a professor of anatomy at the University of Witwatersrand in South Africa. Dr. Lubenow wrote that he "acquired a skull that had come from a lime works at Taung. Dart immediately recognized that the skull was something unique. After cleaning it and studying it, he announced to the world that he had discovered our evolutionary ancestor. It was the skull and endocranial cast of an extinct primate child which Dart named *Australopithecus africanus*." Lubenow went on to say that "until Lucy was discovered in late 1974, Taung, the type specimen of *Australopithecus africanus* was considered our oldest evolutionary ancestor. Although dating the South African fossils has always been a nasty problem, Taung was generally considered to be between two and three million years old. That age seemed appropriate for it as our evolutionary ancestor."[8]

The Taung skull was the first fossil that Dart called *Australopithecus africanus*. Dart considered this fossil to be that of a hominid because of what be perceived to be human-like characteristics. However, the skull of a very young ape does not completely resemble that of an adult ape. This skull was from an immature ape, about three years of age. It was predictable that the Taung skull would not have brow ridges and would have smaller canines. Furthermore, the British anatomist Lord Zuckerman was never convinced that the *australopithecines* walked upright.[9] These facts throw serious doubt on Dart's claim that *A. africanus* was an evolutionary ancestor of mankind.

Dr. Lubenow wrote that "the more serious problem for *africanus* is that fossils identical to those of modern humans parallel the entire history of *africanus*. Thus, *Africanus* cannot be our ancestor."[10] Louis Leakey found that *Homo erectus, Homo habilis*, and *Australopithecus* existed contemporaneously. He found fossils of all three in Bed II during his research at Olduvai Gorge. Furthermore, people were astonished that Leakey found the remains of a round stone shelter at the base of Bed I. The manufacture of these stone huts are not attributed to any hominid, only to *Homo sapiens*.[11]

Australopithecus boisei is another category of the *australopithecines*. It was originally called *Zinjanthropus*. The Leakeys discovered, what they called, *Zinjanthropus boisei* at Olduvai Gorge in Tanzania. This find was really very similar to *A. africanus*. However, the Leakey's research was funded by the National Geographic Society, which wanted a dramatic discovery. So Louis Leakey made some unfounded claims concerning the uniqueness of *Zinjanthropus*. These claims were well publicized in the *National Geographic* magazine. Eventually, Louis Leakey admitted that *Zinjanthropus* was actually an *australopithecine*.[12]

According to Morris and Morris, "...the *australopithecines* were simple apes of some kind is evident from their skulls, which have long been recognized as having the brain capacity (about 500 cubic centimeters) of a true ape. It was long believed, however, that their brains were at least probably human-like in shape. This now also turns out to have been quite wrong." Research on australopithecine endocasts shows that their brains were ape-like in appearance.[13] Dr. Gish wrote that "all of these animals possessed small brains, the cranial capacity averaging 500 c.c. or less, which is in the range of a gorilla, and about one-third of that for man. These animals thus unquestionably had the brains of apes, regardless of what else can be said about them."[14]

Australopithecus africanus is said to have evolved into *Homo habilis*.[15] Dr. Lubenow wrote that "in 1964, Louis Leakey, Phillip Tobias (University of Witwatersrand) and John Napier (University of London) announced in Nature a new human ancestor: *Homo habilis*. Since some of those fossils were found in Bed I, they were also dated at 1.8 m.y.a. From the start, these fossils were the subject of intense controversy. Some felt they were just a mixture of *australopithecine* and *Homo erectus* fossils and hence did not constitute a new taxon."[16] Leakey, Napier, and Tobias felt that these fossils were more advanced than *Australopithecus*, and, therefore, belonged in the genus called *Homo*. This was very controversial, and Leakey had many critics. Even though the cranial capacity of *Homo habilis* was said to be about 650 c.c., many scientists still believed that those fossils belonged in the *Australopithecus* genus. Dr. Gish wrote that "the specimens that Louis Leakey designated *Homo habilis* had been discovered by his team at Olduvai Gorge not long after the discovery of his '*Zinjanthropus*' (*A. boisei*)."[17]

The best known fossil that has been placed in the *Homo habilis* category is usually referred to as skull 1470. The fossilized cranium called skull 1470 was discovered in Kenya, in 1972. This discovery was made by Richard Leakey's team. Leakey initially dated this fossil at nearly three million years old, based on the rocks in which they were found. The cranial capacity of skull 1470 was estimated to be about 800 c.c. Based on the morphology of the skull cap, Leakey believed that this fossil was of the genus *Homo*. He said that the species was indefinite, even though he originally said that skull 1470 may be more advanced than *Homo erectus*, in some respects.

Another fossilized skull, called 1590 has also been placed in the *Homo habilis* category. According to Lubenow, "the disparity of cranial volumes, with 1470 and 1590 being within the human range, while others, such as 1805, 1813, and O.H. 24, are far too small to be considered human." He went on to say that "there is no compelling reason why the large *Homo habilis* material (skulls 1470 and 1590) cannot be classified as *Homo sapi-*

ens based upon their morphology." Skull 1590 was put together from pieces of cranial and dental material. This fossilized skull is about the same size as 1470. However, it appears to have been that of an individual who had not reached full maturity. If this individual had reached full maturity, Skull 1590 would have been even larger. A cranium the size of 1470 is, ostensibly, at the upper end of the *Homo habilis* range. Tlerefore, there is no justification for placing 1590 in the *Homo habilis* category.

Skull 1470 qualifies as a human cranium on the basis of its size, shape, and the thickness of its walls. 1470 also seems to have had a Broca's area, which is that area in the brain of humans that controls the muscles of speech. Skull 1470 was the subject of an article in the June 1973 issue of *National Geographic* magazine. 1470 was given an apelike nose when an artist portrayed the individual from which that skull came as a young female. This female appeared to be quite human, except for the nose. Dr. Lubenow wrote that "human noses are composed of cartilage which normally does not fossilize, and the nose is missing on 1470. It is obvious that the purpose in giving the reconstructed skull 1470 woman an apelike nose was to make her look as 'primitive' as possible. The decision of what kind of nose to give her was an entirely subjective one made by Matternes or his advisors. With a human nose, none would question the full humanity of that woman." Jay Matternes was the artist responsible for that picture in *National Geographic*.

In spite of these characteristics, 1470 was eventually declared to be a *Homo habilis* fossil. Furthermore, the rock formation, in which 1470 was found, was redated to conveniently correspond to a presumed *Homo habilis* time frame. The rocks in which Leakey found skull 1470 were originally dated to be 2.9 million years old. However, in 1981 it was decided that those rocks, as well as skull 1470, were only 1.9 million years old. This age is considered to be about right for a *H. habilis* fossil. Dr. Lubenow went on to say that the "case study of the dating of the KBS Tuff and of skull 1470 offers clear evidence that when the chips are down, factual evidence is prostituted to evolutionary theory."[18]

A partial skeleton that had been put in the *Homo habilis* category is called OH 62. Dr. Gish wrote that "OH 62, classified by Johanson, et.al., and by Hartwig-Schere and Martin as *H. hablis*, was dated at 1.8 million years. According to those investigators OH 62 was a female about three and one half feet tall and was even more apelike than 'Lucy.'"[19] Concerning OH 62, Dr. Lubenow wrote that "it was the first time that postcranial material had been found in unquestioned assocation with a *Homo habilis* skull. The surprise was that the body of this *Homo habilis* adult was not large, as *Homo habilis* was supposed to be. It was actually smaller than Lucy." The point is that OH 62 does not belong in the *Homo habilis* category any more than 1470 does. *H. habilis* seems to be a catch-all classification that contains a mixture of australopithecine and human material.[20]

Homo habilis is said to have evolved into *Homo erectus*.[21] The time frame for *Homo erectus* is said to be from about 400,000 years ago to about 1.6 million years ago. These dates actually represent an average of dates that were given by several evolutionists. Dr. Lubenow wrote, "These dates for *Homo erectus* represent a 'comfort zone' for the evolutionists. It is not difficult to see that these dates position *Homo erectus* in the relevant time period to serve as that much needed transitional taxon that progresses toward modern humans."[22]

There are a number of characteristics that are said to be typical of *Homo erectus* morphology. These include an elongated skull that is relatively low and broad, an unusually thick cranial wall, minimal chin, a remarkably large jaw, large teeth, supraorbital ridges, and thick, heavy bones of the post-cranial skeleton.[23] Java Man was the first to be called *Homo erectus*. Since then, the remains of over 200 individuals have been designated as *Homo erectus*.[24]

The discovery of the remains of Java Man is credited to a Dutch physician named Eugene Dubois. He had been sent to Java in 1889. Originally, just a skull cap was found on the bank of the Solo River in the fall of 1891. Then a year later he is said to have found a femur about 15 meters from where the skull cap had been. Three teeth were found at the same site. Dubois believed that the skull, femur and teeth were the remains of the same individual. He decided that this individual was a true transitional form between apes and man. In other words, Dubois claimed to have found a missing link. He named this alleged ape-man *Pithecanthropus erectus*, which means erect ape-man. Eugene Dubois took the Java Man fossils to the International Congress of Zoology in 1985. This meeting was held in Leyden, the Netherlands. His claims were met with much skepticism.

Dubois, an evolutionist, did not mention the fact that two human skulls were discovered near the place where he had found the Java Man skull cap. The other skulls had cranial capacities which actually exceeded the average capacity of modern humans. Dubois evidently did not reveal this fact because it would have made his claims that Java Man was a missing link less credible. He did not say anything publicly about these two other skulls, called the Wadjak skulls, until 1922.[25]

Dr. Marvin Lubenow is a professor at Christian Heritage College in El Cajon, California. He has researched Java Man for over twenty years. Dr. Lubenow wrote that, "my conclusions on Dubois and Java Man are as follows: (1) Java Man is not our evolutionary ancestor but a true member of the human family, a post-Flood descendent of Adam and a smaller version of Neanderthal; (2) Dubois seriously misinterpreted the Java Man fossils, and there was abundant evidence available to him at that time that he had misinterpreted them; (3) the evolutionists' dating of Java Man at half a million years is highly suspect; (4) more modern looking humans possibly including Wadjak Man were living as contemporaries of Java Man; and (5) Java Man was eventually accepted as our evolutionary ancestor in spite of the evidence because he could be interpreted to promote evolution."[26]

The Wadjak skulls were found at the same level in the strata that the Java Man remains were discovered in.[27] Dr. Lubenow wrote that "to pre-serve the uniqueness of *Pithecanthropus* as the missing link, Dubois had to make sure that no fossils of more modern morphology could be assigned to the same stratigraphic level or given the same date." The accepted date of Wadjak is about 10,000 years ago. The evolutionists do not think that Wadjak Man had any part in human evolution, because Wadjak Man was clearly human. According to Lubenow, "there is evidence that Wadjak was ap-proximately the same age as *Pithecanthropus*, to sell *Pithecanthropus*, Dubois had to hide Wadjak."[28]

It is important to remember that Eugene Dubois was not a professional geologist or paleontologist. It is doubtful that he had the background to determine the age of Java Man. This seems to have been true whether you are talking about an evolutionary context or not. At that time, very little was known about the geology of Java anyway. To complicate matters even further, Dubois was not at the site of excavation very much. The digging was actually done by Javanese laborers under the supervision of two Dutch army corporals. These two soldiers, that were assigned to help Dubois, were from the engineering corps.

Dubois' reports on Java Man were said to be rather sketchy in terms of geological information. Dubois was completely dependent on these two cor-porals who apparently had no training in geology. Careful documentation, con-cerning the exact layers of rocks that fossils are found in, is absolutely vital for dating fossils. Even if you accept the evolutionary time-scale as valid, the dating of Java Man at 500,000 years ago is without an adequate geological foundation.[29]

Besides the problems with the dating of the Java Man fossils, there was the problem of the classification of those fossils. Was Java Man an ape, a man, or an apeman? Many scientists said that the skeletal remains of Java Man resembled that of a very large gibbon. A gibbon is a member of the family *Pongidae*. Gibbons travel very quickly through the forest by swinging from branch to branch. The average adult gibbon is about three feet tall and is very light in weight. The director of the French Institute of Human Paleontology, Marce Boule, said that the Java Man skull was that of a large ape. Boule also said that the Java Man femur was probably that of a human. Eugene Dubois, himself, later said that *Pithecanthropus* re-sembled a very large gibbon in some ways. [30]

The Java Man had been estimated to have a cranial capacity of about 900 to 1,000 c.c. This is within the human range. The average brain size of a gibbon is only about 100 c.c. Even a giant of a gibbon wouldn't have a brain ten times that size. The mighty gorilla has a cranial capacity of less than 600 c.c. The Java Man had a human sized brain. Furthermore, the Java Man femur was virtually identical to that of a human. H. L. Shapiro, among others, said so. The question is, why were all those people trying so hard to make a monkey and/or missing link out of a man?

The answer has to do with that ancient concept called the Great Chain of Being. Dr. Lubenow said that this concept was "patterned not after Moses, but after Plato. According to this concept the Almighty had created a great ladder or chain of living things."[31] The Great Chain of Being can be traced back to Plato. Aristotle took Plato's idea and somewhat modified it.

Dr. Henry Morris wrote that "the millennium in Europe between the fall of Rome and the Renaissance (400 to 1400 A.D.) was dominated, of course, by the Catholic Church and therefore by nominal allegiance to the Biblical/creationist world view. Nevertheless, the evolutionism of the old pagan world had not died; it had merely gone underground, as it were." He also said that "the Renaissance (meaning 'rebirth') has been so named for the very reason that the submerged preChristian culture of Greece and Rome was revived in this period." This submerged culture included the ancient chain of being concept.[32] According to Lubenow, "it is obvious that in the Great Chain of Being we are dealing not with Biblical concepts but with pagan Greek philosophy."

The Chain of Being concept involved small, progressive changes between organisms. Since Darwinism was, largely, based on the Great Chain of Being, it also involves small, progressive changes with no gaps between organisms. The remains of Java Man were first interpreted in this philosophical context. This was also true of the early Neanderthal discoveries.[33] If Darwinism is true, there must be intermediates between the apes and mankind. The problem, as far as the evolutionists were concerned, was that there was a big gap between apes and man. But a chain has no gaps. Darwinism really demands that ape-men existed and, as far as Java Man was concerned, the scientific community was trying to deliver.

According to Morris and Morris, it is likely that "fossils have been identified as *Homo erectus*, rather than *Homo sapiens*, simply because they happened to be members of the human race whose brain sizes were at the low end of the normal spectrum of brain size variation, but otherwise they were probably normal human beings." Morris and Morris wrote that "language, of course — that is, the ability to communicate in abstract, symbolic, intelligible speech, whether verbal or written is the one essential attribute that distinguishes true man from apes or other animals. And, as just noted, *Homo erectus* meets that test, even if his cranial capacity was at the lower end of the human spectrum."[34]

Dr. Henry Morris wrote that "one of the most obvious and unequivocal proofs of the uniqueness of man in contrast to the animals is the ability to communicate in terms of intelligible, abstract, symbolic human language. Animals bark and grunt and chatter, but this attribute is completely and qualitatively different from human speech."[35] According to Meldau, "man's mind alone is created with the ability to learn and use language. Beyond all question, this is one of the greatest gifts God gave to man. Tle ability to learn and use it distinguishes him from all lower animals."[36] The editors of *Encyclopedia Britannica* place the origin of human language at over one million years ago. One million years ago is right in the middle of the *Homo erectus* time frame.[37]

Besides brain size and the use of language, there is another important similarity between *H. erectus* and modern humans. This similarity has to do with reproduction. D.C. Johanson believes that if a *H. erectus* man and a *H. sapiens* woman mated (or vice versa), they could have produced fertile offspring. This ability, to produce fertile offspring, is the basis of the Biblical concept of kinds. Dr. Lubenow wrote that "the fact that Johanson believes that *erectus* and *sapiens* could mate if they were living at the same time is actually a confession that the differences between them are not great. Furthermore, the fossil record shows that *sapiens* and *erectus* were living at the same time. The differences between *Homo erectus* and *Homo sapiens* are the result of genetic variation rather than evolution."[38]

Homo erectus is said to have evolved into the category called archaic *Homo sapiens*. According to Lubenow, "this category includes a minimum of forty-nine fossil individuals who do not fit into either the Neanderthal or the *Homo erectus* categories. The reasons are that (1) they have a somewhat different skull morphology from the Neanderthals, (2) many of this group are dated much earlier than the 'classic' Neanderthals, although more than half of them are Neanderthal contemporaries, and (3) they have a cranial capacity that is too large for them to be classified as *Homo erectus*." The best known fossil in this category is called the Rhodesian Man. It has been suggested that the name of this category be changed to *Homo sapiens rhodesiensis*. The cranial capacity of archaic *Homo sapiens* is said to be from about 1,100 to 1,300 c.c. Members of this category had long faces which were quite large. Their jaws were very prominent. The Rhodesian Man had very heavy ridges over the eyes.[39]

The remains of Rhodesian Man were discovered in Zambia in 1921. This discovery was in what was formerly Northern Rhodesia. The skull morphology of Rhodesian Man was quite unusual. Dr. Lubenow wrote that "because the brow ridges on this fossil are more pronounced than those found on any other human fossil, no human fossil appears to be more 'primitive', 'savage', or 'apelike' than does Rhodesian Man. Yet his brain size of 1280 cc is so large that the fossil demands to be classified as *Homo sapiens*. We need to be constantly reminded that there is nothing in the contours of the skull of an individual that gives clues as to his degree of civilization, culture, or morality."

The remains of Rhodesian Man were found in a mine shaft, about 60 feet beneath the surface. The skull was found along with the skeletal remains of a few other individuals. The upper jaw of one of the others had a very modern morphology. This would indicate a considerable degree of genetic variation within that small assemblage of just a few individuals. The post-cranial bones which were found in that mine shaft were much like modern humans. The remains of Rhodesian Man have been dated, by evolutionists, to be between 40,000 and 400,000 years old. However, it is quite remarkable that the skull was found in a high state of preservation. This skull showed no signs of having been mineralized. It is hard to imagine how a skull that old could not have been mineralized.

The Rhodesian Man supposedly lived in the far distant, prehistoric past. However, Dr. Lubenow wrote that "this individual was either mining lead and zinc himself or was in the mine shaft at a time when lead and zinc were being mined by other humans. This smacks of a rather high degree of civilization and technology." Lubenow also wrote that the "Book of Genesis clearly confirms the advanced culture and technology of the ancients, specifically mentioning metallurgy in Genesis 4:22 and music in Genesis 4:21." The point is that the actual setting of the location where Rhodesian Man was found indicates that he was clearly a man. Furthermore, this man was probably a descendent of Noah. [40]

The archaic *Homo sapiens* are said to have evolved into a group called *Homo sapiens* Neandertal[41] (also Neanderthal). This group, obviously, takes its name from the famous Neanderthal cave-men. The first Neanderthal was discovered in 1856 on property owned by Herr von Beckersdorf. According to Lubenow, "although the bones were obviously human, they looked different. Beckersdorf took the bones to a science teacher, J.K. von Fuhlbrott, who was also president of the local natural history society. Recognizing the antiquity and the extreme ruggedness of the bones, Fuhlbrott felt that they were probably the remains of some poor soul who had been a victim of Noah's Flood."

The Neanderthal bones were also examined by Rudolf Virchow, who was a member of the faculty at the University of Berlin. It was Virchow's opinion that the bones weren't old enough to have been in the Flood. Professor Virchow said that they were from a man who had rickets as a youth, then arthritis when much older. He determined that this individual had somehow received a number of serious blows to the head.

Those Neanderthal remains were eventually examined by an evolutionist named William King, who was a professor at Queen's College in Ireland. He said that these bones were not truly human and must be the remains of a member of a very primitive, subhuman race, which he called *Homo neanderthalensis.* Highly imaginative drawings of a stoop-shouldered, ape-like brute called Neanderthal Man were soon circulated.[42] Dr. Gary Parker wrote that "evolutionists chose to believe in the evolution of man for philosophical or religious reasons, not because of logical inference from the fossil evidence. In his book, *The Descent of Man* Darwin did not cite a single reference to fossils in support of that belief. And there were several fossil specimens of human beings known when Darwin wrote, namely Neanderthal Man."[43]

In 1886, two very similar fossilized skeletons were discovered in a cave near Spy, Belgium. As a matter of fact, the remains of a wooly rhinoceros and a mammoth were discovered in the same cave. These skeletal remains made it clear that the first Neanderthal was a member of a group of ancient humans. The remains of over 300 individuals, classified Neanderthal, have been discovered.[44] Neanderthals were fully human. Dr. John Morris wrote that they "had a cranial capacity which exceeded the average human today. They buried their dead with religious significance, and were probably the source of at least some cave drawings. They crafted and used tools, cultivated crops, and no doubt had a spoken language. Some evidence exists of record-keeping and possible writing. These were totally human individuals." He also wrote that "creationists have long suspected that the Neanderthals were an ethnic group who migrated away from the incident at the Tower of Babel, complete with a language, but perhaps little technology. Soon they found themselves in harsh, European ice age conditions, and were forced to eke out a living with poor nutrition, living in caves." Bones with Neanderthal characteristics have also been found in Asia, Africa, as well as in Israel.[45]

In fact, there has been a very important discovery in Israel concerning the speech of Neanderthals. Some evolutionists have said that Neanderthals were so primitive that they lacked speech. However, a Neanderthal skeleton found in Kebara Cave seems to provide evidence that this allegation is false. A small bone, called the hyoid bone, was found in Kebara cave. The hyoid bone is located at the base of the tongue and is connected to the voice box, or larynx. This small bone is considered to be vital to the capability of speech. This hyoid bone, found as part of the Neanderthal skeleton in Kebara Cave, was nearly identical in morphology to hyoid bones found in modern humans. The Israeli research team concluded that Neanderthals were fully capable of speech.

As previously mentioned, the Neanderthal remains had some peculiar characteristics. Compared to modern humans, a Neanderthal had a skull that would be considered to be elongated, broad, with a rather low forehead. They had large faces which were long, with heavy brow ridges. Their faces appear to have jutted forward in the center. Their chins were rounded and weak. The post cranial (below the skull) skeleton of the Neanderthals was very rugged, with bones which were quite thick. Big bones require big muscles for movement. Even though the Neanderthals did differ from most modern humans, it was a mistake to label them as subhuman. Even the prominent evolutionist, Thomas Huxley, said that the Neanderthals were fully human. Huxley did not believe the Neanderthal Man was an evolutionary ancestor of modern humans. In terms of physical strength, Neanderthals appear to have been far superior to most modern humans. Dr. Trinkaus, a Neanderthal authority, says that their limb and trunk bones were so massive as to suggest a nearly superhuman robustness. The evidence indicates that this was true of both males and females. Their skeletons reflect an exceptionally strong musculature.[46]

There are many evolutionists that would say that the Neanderthals became extinct many thousands of years ago. However, this is really a historical issue rather than a scientific issue. Without the necessary historical/genealogical records (which do not exist), no one can categorically say that the Neanderthals died out 30,000 to 35,000 years ago. In fact, there are authorities who say that modern eastern Europeans are the descendants of the Neanderthals.[47]

There may be evidence of this ancestry related to the enormous physical strength of the Neanderthals. It might not be entirely coincidental that the east Europeans are so successful in international weightlifting competition. This is especially true of the heavier classes. For example, at the 1996 summer Olympics, lifters from eastern Europe won 73% of the medals in the five heaviest weight classes.[48] Most of the rest of the world can't even begin to compete with them in the heavyweight classes. However, they also do well in the lighter classes, At the 1976 Olympics, eastern Europeans won gold medals in every weight class.[49]

This covers the alleged family tree of *Homo sapiens*, as well as the Neanderthal Man who was *Homo sapiens*. This evolutionary sequence began with *Australopithecus afarensis*, which is dated from about 3 to 3.6 million years ago (m.y.a.) Next is *Australopithecus africanus*, which is dated from about 2 to 3 m.y.a. The next link in the chain is said to be *Homo habilis* which is dated from about 1.5 to 2 m.y.a. *H. habilis* is ostensibly followed by *Homo erectus*, which is dated from about 0.5 to 1.5 million years ago. Archaic *Homo sapiens* is said to have appeared nearly half a million years ago.

There is a huge problem with this sequence, besides those already mentioned. Fossilized bones which are just like those of modern humans have been dated at 4.5 m.y.a.[50] This age was determined by the same methods used to date the evolutionary sequence just mentioned. Phillip Johnson wrote in chapter six of *Darwin on Trial*, "Physical anthropology, the study of human origins, is a field that throughout its history has been more heavily influenced by subjective factors than almost any other branch of respectable science. From Darwin's time to the present the 'descent of man' has been a cultural certainty begging for empirical confirmation."[51]

In a sense, Darwinism demands that ape-men have existed. The pressure on scientists to provide confirmation that humans evolved, led to one of the most famous cases of fraud in history. In 1912, Dr. Charles Dawson and Arthur Smith Woodward announced that they had discovered a jawbone and a portion of a skull. These fossils were said to have been discovered in a gravel pit in the vicinity of Piltdown, England. The skull appeared to be similar to that of a human but the jawbone seemed to resemble that of some type of ape. Dawson and Woodward claimed that the jawbone and skull were from the same individual. This alleged ape-man was called the Piltdown Man. The Piltdown Man was said to have lived about half of a million years ago.

However, by 1950 a method had been developed to calculate the relative age of fossil bones. This procedure is based on the amount of fluoride that is absorbed by bones that are in the soil. It was found that the jawbone of Piltdown Man contained almost no fluoride. Therefore, it could not have been a fossil. It turned out that the Piltdown Man was a fake. A careful examination of the bones showed that they had been treated with iron salts which gave them the appearance of great age. It was found that the teeth had been filed to make them appear to be more worn than they actually were.[52]

There is a possibility that the skull of Piltdown Man was actually found in the pit. The Piltdown Common had been a burial ground during the mid-fourteenth century because of the great plague. Dr. Marvin Lubenow wrote, "Both the mandible and the canine tooth were determined conclusively, by collagen reactions, to be those of an orangutan." The teeth of the jawbone had been filed to make them appear less ape-like and the file marks were fairly obvious. At least a dozen people have been accused of responsibility in the Piltdown Man hoax. Who was actually responsible remains a mystery.[53]

Yet, according to Dr. John Morris, "there is no mystery as to how such obviously fraudulently-doctored bones could fool the world's greatest experts for 40 years. They must have wanted desperately to see ape-like characteristics in the human skull, and human-like characteristics in the ape jaw, and they succeeded." The scientific community should have known that something was wrong when an elephant femur, which had been carved to resemble a cricket bat, was found in the Piltdown pit.[54] Fred J. Meldau wrote, "We already have called attention to the colossal fraud perpetrated by, or in the name of Charles Dawson, anthropologist of England, in foisting off on the public (and the world of science as well) the 'Piltdown Man'. True science cannot be blamed for such a forgery, but the whole affair shows that MANY SCIENTISTS CAN EASILY BE DECEIVED."[55]

Another alleged ape-man was called Nebraska Man. The amazing thing about the Nebraska Man was that all that was ever found of him was one tooth. Dr. John Morris wrote, "From a single tooth was drawn a whole family. The naked ape-man, sporting his club, was flanked by his naked wife gathering roots for supper. Behind them were a herd of camels and a herd of horses, whose fossils had been found in the same deposit, but were extinct in that location long ago. The imaginative newspaper coverage and timing of the find made a big impression at the 1925 Scopes Trial."[56] Dr. Raymond Barber, pastor-emeritus of the Worth Baptist Church in Fort Worth, Texas, wrote that "one of the notable hoaxes of the evolutionists is the Nebraska Man, whose fossilized remains were discovered in 1922 by Harold Cook in the state of Nebraska. He was said to have lived a million years ago. Time and God have always been on the side of creationists; therefore, when the real evidence was produced, the so-called fossil was a tooth not of a prehistoric man, but of a pig."[57]

Over the years, some other evolutionists have resorted to deceitful methods in an attempt to persuade the unsuspecting public. Perhaps the best example of this is the arrangement of fossils into a sequence that supposedly provides proof that humans evolved from apes. They will often take a collection of human and primate skulls and arrange them in order of increasing size, and then they will say that this represents an evolutionary progression. Such sequences are really nothing more than scientific sleight of hand and these sequences prove nothing, at all, about human origins.

These fossil sequences usually contain a mixture of *australopithecine* and human material. The *australopithecines* had an average cranial capacity of about 500 c.c. Human skulls have a cranial capacity of about 1500 c.c., but have a broad range from about 700 c.c. to over 2000 c.c. So they start with *australopithecine* skulls and also select human skulls of various sizes. Then all of those skulls are put in a sequence according to size. These sequences prove nothing about evolution since the fossil record provides evidence that *Australopithecus, Homo sapiens*, and *Homo erectus* all lived at the same time.[58]

According to Dr. Lubenow, "some human fossils are arbitrarily downgraded to make them appear to be evolutionary ancestors when they are in fact true humans." Virtually any set of objects, whether created by God or humans, can be organized into a sequence that ostensibly reflects evolution.[59] Dr. Lubenow also wrote that "we now know that relative brain size means very little. The relationship between brain size and body size must be factored in, and the crucial element is not brain size but brain organization. A large gorilla brain is no closer to the human condition than is a small gorilla brain. The human brain varies in size from about 700 c.c. to about 2200 c.c. with no differences in ability or intelligence."[60]

The *australopithecines* were apes. The fact that they might have walked upright proves nothing. Chimpanzees and gibbons can walk upright. The *Homo habilis* category is a false taxon which contains a mixture of human and *australopithecine* fossils. The *Homo erectus* category consists of fossils of humans which were rather small. The Neanderthals were a primitive group of people who used tools, made simple hearths, and respectfully buried their dead. The Piltdown Man and Nebraska Man were hoaxes. The fossil sequences are meaningless.

In conclusion, we find that the fossil record does not provide evidence that humans have evolved. "The concept of evolution," wrote Dr. Huse, "is actually nothing more than a scientist's version of the old nursery tale about a frog being transformed into a prince. In this case, the magical transformation is not instantaneous, but requires the magical wand of geologic time."[61]

CHAPTER 9

LIVING FOSSILS

Besides the many problems with the evolutionary time scale that have already been mentioned, there is the problem of living fossils. These are supposedly extinct creatures that have been found to be alive and well. They were thought to have been extinct because at some point, they ceased to appear in the fossil record. The evolutionists thought that because the fossiliferous strata represent long "ages" of geologic history there was no way an organism could still exist if its fossils did not appear in intervening ages. Those organisms that still exist despite being absent in these supposedly intervening ages are called living fossils.[1]

One example of a living fossil would be the *tuatara* of New Zealand. This creature is said to be the only surviving member of an order of reptiles called beakheads. Why did the *tuatara* survive if all its "relatives" died out millions upon millions of years ago? A fossilized tuatara skeleton has been found in Jurassic rocks that are said to be 150 million years old. If the theory of evolution is valid, then why did this creature not evolve into something else by now?

Another example of a living fossil is a type of segmented mollusk that was found in the Acapulco Trench. This type of mollusk, called *Neopilina*, was said to have been extinct since the Devonian period, which supposedly ended about 280 million years ago. Whitcomb and Morris wrote that "280 million years is a long time and one cannot help but wonder about its reality. Fossils of this class of mollusk were apparently plentiful in the early Paleozoic strata and it is amazing that none have been found in the marine strata of the Mesozoic or Tertiary, if indeed these actually represent the hundreds of millions of years following the Paleozoic that they are supposed to." Evolutionists don't really have a convincing explanation for the existence of this living fossil.

An example from the plant kingdom is that of the metasequoia. These trees were found in an isolated area in China in 1945. Metasequoia is actually a genus of conifers that is thought to have had a wide distribution in the northern hemisphere but to have been extinct for about 20 million years. Fossilized specimens of this genus have been found in Eocene and Miocene strata. A paleobotanist named Chaney found dozens of these trees thriving. One of them was close to 100 feet in height. Pliocene and Pleistocene strata do not reflect the continued existence of these trees. If you accept the standard geologic time scale as valid, then the survival of the metasequoia is an anomaly that is impossible to explain. However, when viewed within a Biblical context the existence of metasequoia is not anomalous at all.[2]

As previously mentioned, creationists believe that the fossiliferous strata were deposited as a consequence of the great Flood of Genesis 7 and 8. These layers of strata were actually deposited over a relatively short period of time. Henry Morris wrote that "the entire worldwide sedimentary column was formed rapidly and contemporaneously exactly as in the Noahic Flood recorded in the Bible."[3] At some point during this Flood, all of the metasequoia trees had apparently been buried by sediments. But some of the seeds managed to survive, probably by floating, This explains why the metasequoia trees are still alive. This is one of many examples by which a literal interpretation of the Bible provides a simple explanation to what is a mystery to the evolutionists.

The most famous living fossil is probably the coelacanth. Fossils of this fish are found in strata which was supposedly deposited in the Mesozoic and Paleozoic eras. The coelacanth, (pronounced seel-uh-kanth), is a large fish which weighs about 150 pounds. The initial recorded catch of it was in the Indian Ocean, off the coast of South Africa. Subsequent catches have been in the vicinity of the Comoro Islands. These islands are located between Madagascar and the mainland of Africa. Prior to 1938, the coelacanth was thought to have been extinct for over 70 million years.[4]

The horseshoe crab is another example of a living fossil. The best known species is the *Limulus polyphemus* which inhabits the east coast of this country. Horseshoe crabs are thought to be the only survivors of an extensive group of invertebrates which are called *Xephoswrida*. Members of this group are observed in Silurian strata which is supposedly over 400 million years old.[5] If the evolutionary time scale is valid, then the horseshoe crab has existed, essentially unchanged for hundreds of millions of years.

Another living fossil appears to be an Ordovician index fossil called a graptolite. Graptolites were small marine organisms that probably formed floating colonies. These organisms were thought to have become extinct at the end of the Mississippian period. According to Morris and Morris, "an 'index fossil' is one that is believed to be so identified with a specific geologic 'age' that its presence in a rock is generally believed to date the rock. Graptolites have long been used to date Ordovician rocks, but now it seems that they must also have been present in all other 'ages.'" Graptolites have been found in the ocean near the island of New Caledonia.[6]

They also wrote, "When we consider the vertebrates (fish, amphibians, reptiles, birds, and mammals), once again we find an abundance of living fossils. Among the fish, fossil beds of sharks, herring, catfish, lung-fish and others - as well as whales, dolphins, and other marine mammals - not to mention the famous coelacanth, are abundant. The same is true of land vertebrates."[7] David Raup, a paleontologist at the University of Chicago, concedes that, with the discovery of more fossils, the major groups of organisms keep appearing at earlier dates. Furthermore, Stephen Jay Gould says, basically, that the first appearances of the great majority of phyla of marine invertebrates, that are now living, actually occur no later than middle Cambrian. These findings, by leading evolutionary scientists, are totally consistent with creationism. In other words, Raup and Gould are corroborating the Biblical account of creation in which God concluded the creation after the sixth day.[8] The Bible (as dictated to Moses) says in Genesis 2:1, 2 that "thus the heavens and the earth were finished and all the host of them And on the seventh day God ended his work which he had made; and he rested on the seventh day from all the work which he had made."

When one considers all these examples of living fossils, it seems possible that even some dinosaurs may still exist. In fact, creationists believe that dinosaurs co-existed with mankind for thousands of years. It is probably a consensus among creationists that dinosaurs became extinct in the Middle Ages. However, some dinosaurs may have managed to survive in isolated areas. This seems ridiculous if you accept the evolutionary time scale as valid. A literal interpretation of Scripture indicates that the age of the earth is about 6,000 years. If the age of the earth is actually only thousands of years old, the survival of some of these large reptiles is not unreasonable at all. For example, a creature which seems to have been a plesiosaur was caught in a fishing net near New Zealand in 1977.[9] Plesiosaurs were large marine reptiles that have been referred to as marine dinosaurs.[10]

There have also been reports of creatures that resemble sauropods by the indigenous people of the Congo basin. Villagers in Gabon, in 1979, were shown pictures of various dinosaurs. They identified a picture of a *Diplodocus* dinosaur as a *N'yamala*. *Diplodocus* was a plant-eating sauropod. Sauropod fossils have been found in Jurassic and Cretaceous strata.[11] The *N'yamala* is said to inhabit lakes that are very deep in the jungle and it is reported to be highly aggressive. It is remarkable that most of Gabon is still unexplored. The swampy jungles of that area present a formidable barrier to the outside world.[12]

Pygmies of the northern Congo claim that an animal resembling an *Apatosaurus* lives in a vast swamp which is a part of the Ndoki region. *Apatosaurus* is also a sauropod. The pygmies call this creature *Mokele Mbembe*. Some of them were shown pictures of various animals and they selected the *Apatosaurus* as most resembling *Mokele Mbembe*. These pygmies said that *Mokele Mbembe* is larger than an elephant.

There have been expeditions to this region to try to find this creature. They have been able to stay for only short periods of time due to problems with weather, disease, and limited visas. Another difficulty would be the navigational nightmare that this forbidding, swampy jungle covering over 7,000,000 acres would present.[13] It would be very easy for a dinosaur hunting expedition to disappear without a trace in such a location.

There are many dangers and difficulties in the Congo basin. Snakes are abundant, often lying on the branches of trees. Waterways are infested with crocodiles. Tsetse flies cause sleeping sickness. Great swarms of gnats and mosquitoes move about as horrid noxious clouds. This equatorial region has an annual rainfall of up to 90 inches per year. The moisture plus heat results in the rapid growth of vegetation. Trees may grow to be quite tall with their lowest branches about 60 feet above the ground. Smaller trees form a dense secondary canopy. Underneath this secondary canopy, bushes and ferns can become quite large.[14] The point is that an expedition to the Congo basin would often have great difficulty penetrating the jungle.

Finding what is, most likely, a small group of animals by aerial reconnaissance would not be such a simple matter either. Searching for them by air in a 7,000,000 acre swamp would be like looking for the proverbial needle in a haystack. Furthermore, the dense vegetative canopy would provide very effective concealment for anything beneath.

Dr. Roy P. Mackal is a zoologist at the University of Chicago who has investigated the reports of *Mokele Mbembe*. According to Jerome Clark, who is a leading expert on UFOs and other anomalous phenomena, "Mackal led two expeditions to the area, the first in the company of herpetologist, James H. Powell, Jr., who had heard mokele mbembe stories while doing crocodile research in west-central Africa. Neither expedition produced a sighting, though Mackal and his companions interviewed a number of native witnesses. The creatures, greatly feared, were said to live in the swamps and rivers." One of these creatures was said to have been killed by pygmies at Lake Tele, in the Congo in 1959. The expeditions headed by Mackal were unable to reach Lake Tele, which is said to be virtually inaccessible. However, an expedition led by an American, named Herman Regusters, did manage to reach Tele in 1981. He reported many sightings of enormous sauropods. Mr. Regusters said that camera problems prevented him from photographing these creatures.

There is a long history of accounts of dinosaur-like animals in Africa. A narrative of French missionaries in the west-central part of Africa was published in 1776. This document records strange tracks, shaped like plates, that were very large. These tracks are believed to have been made by the creature that was later called *Mokele-Mbembe*. Mr. Clark wrote that "in the next two centuries missionaries, colonial authorities, hunters, explorers, and natives would provide strikingly consistent descriptions of the animals supposedly responsible for tracks of this kind."[15]

Dr. Bernard Heuvelmans provides extensive documentation of sightings of dinosaur-like creatures in his book, *On the Track of Unknown Animals*. In this book, which was originally written in French, Dr. Heuvelmans quoted Dr. Leo von Boxberger, a German magistrate in Africa. In 1938 he said that "I collected a variety of data from the natives about the mysterious water-beast, but, alas, all my notes and also the local description of the animal were lost in Spanish Guinea when the Pangwe attacked the caravan carrying my few belongings. All that I can report is the name mbokalemuembe given to the animal in Southern Camaroons... The belief in a gigantic water-animal described as a reptile with a long thin neck, exists among the natives throughout the Southern Camaroons, wherever they form part of the Congo basin and also to the west of this area, doubtless wherever the great rivers are broad and deep and are flanked by virgin forest."[16]

The large sauropod-like reptiles that are said to inhabit the Congo basin actually go by a variety of names. *N'yamala* has already been mentioned and it seems to be nearly identical to mokele-mbembe. There are reports of a creature called *badigui* to the north of the Congo. There are also accounts of an animal called *isiququmadeva* to the south of the Congo. Dr. Heuvelmans wrote that "the African dragon's dossier still contains the legends and reports about large amphibious animals with very long necks, called *mokele-mbembe* (or *mbokalemuembe*) and *badigui* to the north of the Belgian Congo and *isiququmadeva* (perhaps also *chipekwe* and *nsanga*) to the south of that country. They are the kernel of the problem. They alone are relevant to any discussion of the possible survival of a dinosaur in Africa."[17]

Bernard Heuvelmans, who clearly accepts evolutionary assumptions as valid, also wrote that "all we can really say is that there seems to be a large and unknown reptile in Central Africa. Its identity remains to be determined, and in view of its size it is probably, like the crocodiles, the Komodo dragon and the giant tortoise, a relic of the great reptile empire that flourished in the Jurassic period. I have already mentioned the remarkable geological and climatic stability in Africa, which could hardly be better suited for preserving such an ancient type." In other words, Heuvelmans is saying that the equatorial climate of that region is similar to the climatic conditions that existed when the dinosaurs were so numerous. Creationists believe that these conditions existed globally just a few thousand years ago. These conditions existed before the great Flood that is described in the seventh and eighth chapters of Genesis.

Dr. Heuvelmans believes that a comparison of accounts concerning *mokele-mbembe*, *badigui* and *isiququmadeva* shows that they all are describing the same kind of animal. A composite of these accounts indicates that this creature is a sizable quadruped that has a rather small head, which is somewhat triangular in shape. There is, perhaps, a small horn on the end of the head which may not be characteristic of both sexes. The neck is said to be long and flexible. The legs are on the side of the trunk, which is quite thick. The trunk is larger than that of a hippopotamus. The long tail is described as being powerful. This creature's skin is brownish-gray and smooth. Its full length is estimated to be between seven and nine meters. The smoothness of the skin indicates that the scales are small.

When Dr. Heuvelmans calls this creature amphibious he means that it is in the water more often than not. This animal has a vegetable diet and it searches for food on land. It seems to have a strong territorial instinct which causes it to attack, what it considers to be, intruders into its domain. According to some reports, it makes a sort of den in the bank of a river or a lake.[18]

The Bible refers to dinosaurs in the book of Job. In Job 40:15-41:34, we find references to two quite different creatures that are called "behemoth" and "leviathan." In Job 40:15-18 we read, "Behold now behemoth, which I made with thee; he eateth grass as an ox. Lo now, his strength is in his loins, and his force is in the navel of his belly. He moveth his tail like a cedar: the sinews of his stones are wrapped together. His bones are like bars of iron." So the behemoth seems to have been a very large plant eating dinosaur with some type of heavy bony plating for protection.

Contemporary Bible commentators have tried to make people think that the "behemoth" and "leviathan" are creatures that are now living. However, a close examination of this passage shows that these creatures are large reptiles that are normally associated with the Jurassic period. Dr. Henry Morris wrote, "Modern Bible scholars, for the most part, have become so conditioned to think in terms of the long ages of evolutionary geology that it never occurs to them that mankind once lived in the same world with the great animals that are now found only as fossils." Some have said that leviathan was a crocodile but the animal described in Job 41 was not a crocodile. What is described is a huge aquatic animal that resembles the fire breathing dragons of ancient folklore.[19] Leviathan was evidently a marine reptile much like a plesiosaur. Leviathan is also mentioned in Isaiah 27:1.

It should be mentioned that all the ancient nations had legends about dragons. The creatures that they called dragons are almost always quite similar to some type of dinosaur. According to Gish, "Stories of dragons come from people all over the world, not from just a few isolated places. The tales come from the oldest of traditions and history." It is difficult to imagine how so many different cultures in so many locations could have devised such similar stories if the dragons never really existed. It would probably be fair to say that dragon is an archaic term for dinosaur.

In most of these legends, dragons were symbolic of evil. For example, there is the story of St. George and the dragon, which has been immortalized in many great works of art. Saint George was a man who actually lived in the third century AD. History records that he was martyred because of his faith in Christ in the year 303. His memory was held in high esteem by the crusaders. We cannot know for sure whether or not Saint George really killed that dragon. However, he is remembered for being a very brave man.[20]

What is called the legend of Saint George could have been an actual account of an encounter between a man and a dinosaur. There are many such accounts. Ham, Snelling, and Wieland wrote that "even in the tenth century, an Irishman wrote of his encounter with what seems to have been a stegosaurus. In the 1500s, a European scientific book, *Historia Animalsiem* listed several animals, which to us are dinosaurs, as still alive. A well known naturalist of the time, Ulysses Aldrovandus, recorded an encounter between a peasant named Baptista and a dragon, whose description fits that of the dinosaur, Tanystrophecus. The encounter was on May 13, 1672 near Bologna, Italy, and the peasant killed the dragon. So the evidence for the existence of dinosaurs during recorded human history is strong.[21]

It seems that paleontologists cannot bring themselves to admit that dinosaurs resembled dragons so closely. Perhaps they are afraid to do so. According to Morris and Morris, "dinosaurs have been so glamorized as denizens of the long-ago, prehistoric past that they have almost become synonymous with evolutionism in the public mind and evolutionists feel emotionally attached to them, afraid that if they yield on this point, the whole evolutionary structure will crumble."[22] A creature that was quite similar to, if not exactly the same as, a plesiosaur was caught in a fishing net near New Zealand. Evolutionists claim that plesiosaurs have been extinct since the Mesozoic era. The fossilized remains of plesiosaurs are found in Cretaceous and Jurassic rock strata. These fossils can be found around the world. The fossilized plesiosaurs found in Cretaceous rocks are often over 40 feet in length but those found in Jurassic rocks are much smaller.[23]

In 1977, the crew of a Japanese fishing boat pulled up their nets and found something they had never hauled up in their nets before. A very strange creature, that was already dead, had been caught at a depth of more than 900 feet. This remarkable catch had an awful smell and must have been dead for some time. Its length was about 32 feet and its weight was over 4000 pounds. This creature's advanced state of decomposition seems to have been the reason that the fishermen threw it back into the ocean. This was very upsetting to some Japanese scientists who came to study ture. However, the crew did take samples of this strange creature's tissue, made measurements, and also took a photograph. Based on the available evidence, at least one of the Japanese scientists seemed to be convinced that this was, in fact, a plesiosaur.[24]

Daniel Cohen, who was formerly managing editor at *Science Digest*, wrote that in 1977, "newspapers all over the world printed stories of how a Japanese fishing boat had hauled up a decaying 'sea monster' off the coast of New Zealand."[25] He went on to say that "whatever it was had a small head, long tail, and four flippers. The total length was somewhat over thirty feet. It was also so badly decayed that the rotting flesh was literally falling off the bones, and the mass dripped. a fatty white ooze on the deck [and] stank horribly." The sketches of this amazing catch were made by an executive of a fishing company whose name was Michihiko Yano. Mr. Yano drew a creature that was quite similar to a plesiosaur in both size and shape. Although he is not a creationist, Mr. Cohen does acknowledge that it might have been a plesiosaur.[26]

Many people believe that the so-called monster of Loch Ness is a type of living fossil. Most of the descriptions concerning this creature closely resemble that of a plesiosaur. Loch Ness is a very deep lake that is nearly 25 miles long. This lake, which is located in northern Scotland, measures about 900 feet at its deepest point. Waterways connect this lake to the North Sea. The plesiosaur-like creature that has frequently been sighted is usually called Nessie. There are scientists who think that Nessie and other such creatures go back and forth between Loch Ness and the North Sea, feeding on migratory fish. It is difficult to see anything that is beneath the surface of the lake because of the small particles of peat which are evenly dispersed in the water.

An underwater photograph was taken in Loch Ness in 1975, that seems to have been of Nessie-like creature. When developed, the picture was rather fuzzy but still showed an Elasmosaurus which is a type of plesiosaur.[27] The fossil record indicates that an Elasmosaurus could be over 40 feet long, with a very long neck. The head of this creature was small and it had sharp teeth with which to catch fish. The great majority of the marine reptiles probably died during the Flood of Genesis 7 and 8. This was due to the tremendous turbulence and also to the presence of a very large amount of sediment in the water. However, it seems that some of the marine reptiles called Elasmosaurus may have survived. This creature could be the "sea monster" that seafaring men of old claimed to have seen.[28]

The first report of the Loch Ness monster is believed to have been in AD 565. In that year, there was a report of a man that was killed while swimming in the River Ness, which connects Loch Ness to the sea. People on the scene said that a monster had killed the man. According to tradition, this creature fled when Saint Columba made the sign of the cross. A number of rather nebulous references to a large creature in Loch Ness have been documented over the many centuries since then.

However, in the late nineteenth century there were a number of reports which were more definite. Jerome Clark wrote that "in 1879 and 1880 two groups of witnesses saw, if their retrospective testimony is to be credited, a large elephant-gray animal with a small head at the end of a long neck as it 'waddled' from land into the water. Almost 20 years earlier there was a report of what, from some distance, looked like a boat that had been capsized which was plowing through the water. There have been literally thousands of reports since then that seem to be referring to the same thing. There were an especially large number of sightings in the 1930s, most of which followed the improvement of a road on the north side of Lake Ness. For example, a remarkable photograph of Nessie was taken by R. K. Wilson in 1934. It definitely showed a creature that looked like a plesiosaur. Another example would be that of a sighting in 1930. In July of 1930 three fishermen reported that a very unusual living creature had come within about 300 yards of their boat and then turned to move away while they were on Lake Ness. They described this animal as being approximately 20 feet long, and said that it stood about 3 feet out of the water. These fishermen estimated that the creature was moving at a speed of about 15 knots. They claimed that, even at that distance, it created waves that caused a drastic rocking of their boat. Mr. Clark, who is vice president of the Hynek Center for UFO Studies, wrote that "on the afternoon of April 14, 1933, near Abriachan, a village on the northwest side of Loch Ness, a couple in a passing car spotted a mass of surging water. They stopped and over the next few minutes watched an 'enormous animal rolling and plunging' out on the loch." In 1937, eight people claimed to have seen three of these creatures swimming side by side on the lake's surface. The two on both sides were said to be much smaller than the one in the middle which suggests a breeding population in Lake Ness.[29]

The accounts of the mysterious creature that lives in Lake Ness, in most cases, seem to describe a plesiosaur. According to Clark, "the bulk of Nessie reports are of a long-necked creature which those who know something of paleontology think resembles nothing so much as a plesiosaur, an aquatic reptile which according to conventional wisdom vanished into extinction 65 million years ago." Mr. Clark also wrote that "the case for the Ness phenomenon begins, like the case for the UFO phenomenon, with a large body of eyewitness testimony from individuals whose honesty and mental health do not appear to be open to question. Such testimony deserves a respectful hearing and ought not to be dismissed out of hand. Beyond that, the evidence for Nessie includes photographs, films, and sonar traces."

Sonar, for example, has indicated that these creatures are about 20 to 30 feet long. The first reported detection by sonar was in 1954 when a large "Nessie-like" object was detected at 480 feet below the surface. D.G. Tucker conducted sonar studies of Loch Ness between 1968 and 1970. He was a professor at the University of Birmingham There were numerous incidents in which his team tracked creatures that were about 20 feet long swimming deep in Loch Ness. During one such incident they tracked a group that included at least five individuals. Tucker concluded that these were not fish based on their actions, size and speed.

In June of 1975 researchers associated with the Academy of Applied Science, which is based in Massachusetts, took two pictures which are strong evidence in favor of Nessie's existence. The first of these showed a living creature's head, neck, and upper trunk. The other picture, taken seven hours later, showed the face and head at close range. In fact, the head was probably less than five feet from the camera. It was just as a great number of witnesses had described it, appearing to be rather horselike. Both of these photographs were taken underwater. Mr. Clark wrote that "on the strength of the pictures zoologists from the Smithsonian Institution, the Royal Ontario Museum, Harvard University, the New England Aquarium and other prestigious establishments either endorsed Nessie's existence outright or declared it now to be a distinct possibility."

The scientists from the Academy of Applied Science had previously captured this creature on film in 1972. These pictures were taken with an underwater camera that was used in conjunction with a sonar device. The film from this underwater camera was taken to Eastman Kodak's head office to be developed. Two frames of film were remarkable in that they showed a large flipper that was attached to a body. It was in the exact spot where the sonar echoes had indicated. It is true that these images were rather murky due to the peat which is suspended in the water of Loch Ness. So these researchers took the pictures to the Jet Propulsion Laboratory. At the J.P.L., photo analysis is done for the military, the government, and various scientific agencies. A computer enhancement procedure removed much of the murky quality of the pictures. This technique had previously been used successfully in a number of fields including medicine, forensics, as well as the space program. Analysts concluded that the flippers were about four to six feet long.[30] These findings indicate that a living fossil probably resides in Loch Ness.

Besides sauropods and plesiosaurs, other large reptiles that are associated with the Jurassic period may have survived up to modern times. These would be the flying reptiles. The 26 April, 1890 issue of the *Tombstone Epitaph* carried an article which gave an amazing account about two men who encountered a very large flying creature in the Arizona desert. According to this newspaper article, they were riding through the desert when they saw this creature preparing to land. It was said to have been a very frightening sight to both men and horses. The cowboys described the creature as having enormous wings, large claws, and a long, narrow trunk. These claws were on the feet and also on the front of the wings. These men said that the head, by itself, was about eight feet long and that the mouth was full of teeth. The cowboys killed it and cut off a part of its wing. The wing was described as being a strong membrane with a smooth surface, much like that of a bat.

This description would fit that of a member of the pterodactyloid group. It may have been a *Quetzalcoatlus*, which was the largest animal that was ever known to fly. Such a creature reminds one of the Thunderbird of Native American tradition. Many tribes, ranging in location from Mexico to Alaska, have legends of a huge flying beast. The sound of the Thunderbird's wings were said to be like thunder.[31] A fossilized *Quetzalcoatlus* was found in Big Bend Park, in Texas, in 1972. This flying reptile had a wingspread of about 48 feet.[32] It seems possible that those two cowboys killed the last surviving Thunderbird.

There have also been relatively recent reports of flying reptiles in Africa. Bernard Heuvelmans wrote that "in 1923 Frank H. Melland published an account of his travels entitled *In Witchbound Africa*. In one chapter he relates that he kept coming upon rather vague rumors about a much-feared animal called 'kongamato' said to live in the Jiundu swamps in the northwest corner of Northern Rhodesia, near the frontier of the Belgian Congo and Angola. He asked what it was. The natives told him that it was a bird, but not exactly a bird, more like a lizard with wings of skin like a bat's." They later identified a picture of a pterodactyl as the "kongamato." Most modern paleontologists do not believe that it is possible that some flying reptiles survived until relatively recent times. This is because fossils of these creatures are not found after Cretaceous strata. However, this is not absolute proof.[33]

There have been accounts of sightings of flying reptiles in other parts of Africa. For example, there are stories about "flying dragons" that lived in the vicinity of Mount Kilimanjaro until fairly recent times. Furthermore, Percival reported that the indigenous people near Mount Kenya told of a flying beast that was quite large. These people were surely familiar with birds and bats. If this flying beast had been either, they would have designated it as a bird or a bat. The question that must be asked is, how did these primitive African people come to believe in the existence of a large bat-like flying reptile if it had not survived until fairly recently? After all, these Africans knew nothing about the supposed Jurassic period.

The indigenous people of the Camaroons speak of a strange flying creature called *olitau*, which seems to be quite similar to "kongamato." Ivan Sanderson may have actually seen this creature when he was leading the Percy Sladen expedition. He had to dodge a very large, black bat-like creature twice when it was attacking. Sanderson was able to see its sharp, evenly spaced white teeth as its lower jaw hung open. Based on Sanderson's account, Dr. Heuvelman's gave a number of reasons why the flying creature, that attacked Ivan Sanderson, could not have been any member of Megachiroptera. This group includes the largest bats. It should be mentioned that Heuvelmans earned a doctorate in zoology at the Free University of Brussels.[34] There is a strong possibility that Sanderson was attacked by a flying reptile. Furthermore, Dr. Mackal reported that there was a sighting of a pterodactyl-like animal, in a swampy area in Namibia, in 1975.[35]

There are several reasons why the flying dinosaurs, as well as other types of dinosaurs, either disappeared or nearly disappeared. The fossil record reflects that dinosaurs were very numerous in the past. So something drastic must have happened. According to Ham, Snelling, and Wieland, "as far as most people are concerned, they died out about 65 million years before man appeared on earth. This view of the dinosaurs is intimately associated with evolution which is dogmatic about how life arose by chance, yet surprisingly vague about why creatures such as the dinosaurs died out. Citing every possible cause from massive attacks of diarrhea, to meteorite bombardment, or drug addiction as new plant forms evolved, the theories of why the dinosaurs became extinct are often hilarious for their ingenuity."[36]

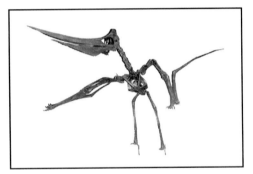

PTERANODON Stenbergi

The real reason for the disappearance of the great numbers of dinosaurs was the climate change that resulted from the Flood. Dr. Gish wrote that "when the dinosaurs left the Ark after the Flood, they found a world incredibly different than the one before the Flood. Food, once abundant, would have been difficult to find. Weather that had always been pleasant would now often be hostile." In fact, the fossil record reflects that Greenland had a sub-tropical climate at one time. The type of fossilized plants and animals that are found in Greenland now live only in places like Puerto Rico. This is also true of fossils found in Antarctica as well as above the Arctic Circle in the north. This indicates that, in the past, the world had a uniformly warm climate.

A scientist named Alvarez has theorized that the dinosaurs became extinct as a consequence of a major catastrophe. He suggests that a huge object, about five miles across, struck the earth and triggered a drastic climatic change. This has become a popular theory among evolutionists. D.T. Gish wrote that "some scientists have suggested the idea that perhaps a huge asteroid struck the earth. Supposedly this monumental collision threw three billion tons of dust into the air, blocking out the sun for several years and causing most plants to die, and so the dinosaurs died out for lack of food. Other scientists argue strongly against this idea. If this were true, why did not all the birds die too? How did all the small, thin-skinned mammals manage to survive? Such a great catastrophe all over the earth would have also killed all the birds, mammals such as dogs, bears, rats, mice and many other animals. How did such reptiles as snakes, turtles, and crocodiles survive? This scenario does not sound at all reasonable."[37]

Ham, Snelling, and Wieland wrote that "any theory about what happened to the dinosaurs, based on bones alone, is merely a scientist's opinion. He was not an observer of the event, and he doesn't have the past against which to test his theory. Thus, such study is not even scientific in the normal sense. It is really his belief about the past, an attempt to explain the evidence which exists in the present. The only way anyone could be sure of what did happen in the past would be through the testimony of any reliable witness who was there. And that is what the book of Genesis is." In this case, the eyewitness is none other than God. The record of Genesis provides a framework, or perhaps context, within which the available evidence actually makes sense.

God tells us in Genesis about the Flood that covered the entire world. This great cataclysmic deluge surely caused the drowning of the dinosaurs. These creatures were consequently rapidly buried by the sediments of the Flood. Ham, Snelling and Wieland also wrote that "most of the dinosaur fossils we find around the earth today are probably all that remain of those killed at the time of Noah's Flood. The contorted shapes of these animals, the massive numbers of them in fossil graveyards, their wide distribution, and the presence of whole skeletons which show convincing evidence of being rapidly buried, all testify to massive flooding." The dinosaurs taken on the Ark would, obviously, have survived the Flood. We know from Genesis 6:19 that dinosaurs must have been on Noah's Ark. As previously mentioned, before the Flood the world was surrounded by a water canopy. After the Flood this canopy was gone and the world's climate was drastically changed. Without this canopy to create a greenhouse effect, our planet began to have drastic fluctuations in weather and temperature. The lush vegetation that the fossil record indicates existed before the Flood, was no longer available everywhere. More radiation was reaching the surface. The world had become a harsher place to live in.[38]

In conclusion, we find that the existence of living fossils presents a big problem to the evolutionists. They generally try to make people think that all the facts fit into a nice, neat package. However, the present-day existence of creatures such as the tuatara lizard, the horseshoe crab, the coelacanth, and colonial graptolites is proof that this is not, at all, true. The existence of living fossils tends to corroborate creationism since the stability of the living fossils, over long ages of geologic history, seems unlikely. If evolution really happens then why did these living fossils not evolve into something else? If evolutionary assumptions are correct the tuatara lizard, for example, has somehow resisted evolution for over 150 million years.

Coelacanth

Galopagos Tortoise

CHAPTER 10

THE EARTH IS YOUNG

The concept of a young earth is rejected by evolutionists, who claim that the earth must be millions upon millions of years old. They make this claim because evolution requires immense spans of time to take place. Lord Kelvin, who was a contemporary of Charles Darwin, calculated that the maximum age of the earth was about 15 million years. Even though this is an extremely long time, this was very disturbing to Darwin. This was because he did not think it was nearly enough time for evolution to take place. In this chapter a young earth means that the age of this planet is thousands, rather than multiplied millions, of years old.

Charles Darwin was determined to promote a philosophical viewpoint. Dr. Chittick wrote that "Darwin held firmly to evolutionary views in spite of contrary evidence. The fossil record did not show the required transitional forms, physical evidence indicated the earth was not as old as his theory would demand. This troubled Darwin, but he chose to disregard the physical evidence so that he could retain a philosophical viewpoint which allowed him to get around Biblical history. Vast amounts of time are a philosophical necessity for those who reject Biblical history."[1]

Charles Darwin has been called an "apostate divinity student."[2] He needed Lyell's uniformitarianism to undermine the historical accuracy of Genesis. A non-catastrophist interpretation of earth history was needed to do this. After all, the concept of a global Flood that drastically changed the geologic characteristics of the Earth is definitely a catastrophist concept. Insofar as uniformitarianism excluded catastrophes as having a major impact on the world, it also excluded the Flood of Genesis as having an important impact. In order to undermine the concept of Biblical catastrophism, Charles Lyell had to undermine all forms of catastrophism.

Biblical catastrophism, which concerns the Genesis Flood, and uniformitarianism are completely contradictory. Uniformitarianism has also been called ages-geology. Dudley J. Whitney wrote that the "belief in aqueous cataclysm, the Noachian Deluge, has come down to us from our earliest ancestors among nearly all races of men, and this seems to provide the only real alternative to ages-geology. Also, evidences are abundant of much aqueous cataclysm in the burial of fossils and in the making of the sedimentary rocks generally."[3] Mr. Whitney, who was a pioneer of modern creationism, wrote a book titled *The Face of the Deep* which contains many amazingly astute observations concerning Earth history.

Radiometric Dating Methods

Radiometric dating procedures are used as a justification for the claims that the Earth is immensely old. According to Chittick, "the idea of vast amounts of time was essential for Darwinism to take hold. Many different geological systems and processes have been suggested for determining the age of the earth. However, in Darwin's day, only those methods Lord Kelvin used were based on solid physical principles. Kelvin's calculations, however, did not give an earth nearly as old as needed by the evolutionist." Dr. Chittick also wrote that "the phenomenon of radioactivity was subsequently used in an attempt to date the earth. With appropriate evolutionary assumptions, it was possible to use data from radioactivity to calculate an age very much greater than that given by Kelvin's calculations. This dating technique could give the vast ages needed by evolutionists."[4]

Radioactive dating methods are said to provide proof that the earth is millions upon millions of years old. Three of these dating procedures are the uranium/lead method, the potassium/argon method, and the rubidium/strontium method. In these methods, what is called a parent element (e.g. uranium), changes, or decays into what is called a daughter element (e.g. lead) at a known rate. The basic rationale of radiometric dating is that when the relative amounts of the parent and daughter elements are measured the age of an igneous rock can be calculated.

A number of assumptions which cannot be proven are involved in the standard radiometric methods of dating rocks. One of these assumptions is that daughter elements are not present in a mineral when it is formed. However, this is almost always an unwarranted assumption. In fact, virtually all minerals of igneous origin do contain some of the daughter product after cooling. This has been found to be true for newly formed igneous rocks, which resulted from the cooling of lava that had been emitted from modern volcanoes. Therefore, the age that is determined by a radiometric dating procedure is typically far greater than the actual age.[5]

Dr. Chittick asked the following questions, "What were the conditions when the earth began? What were the isotope ratios? How much of a radioactive element was present in the beginning? No modern scientist was around to record those initial conditions. We really do not know what they were. All a scientist can do is assume, estimate, or guess what the original composition of any radioactive system was originally."[6] There is a uranium/lead procedure which assumes that a rock sample originally contained equal amounts of parent and daughter elements. However, this assumption is also not able to be proven.[7]

Another assumption which is always involved in radiometric dating is that decay rates have remained constant over time. However, as Dr. Chittick points out, "the phenomenon of radioactivity was only discovered in 1896. Thus, rates of decay of radioactive systems (half-lives) have only been measured for less than a hundred years. Any statement about decay rates in the far distant past has to be extrapolated from present conditions. Extrapolations from a hundred years to millions or billions of years are suspect. It is scientifically unsound to make extrapolations of such magnitude." He also wrote that "the radioactive dating system does not meet the fundamental requirements of a clocking system. We do not know the initial conditions, and even if we did, we cannot be certain that rates of radioactive decay have always remained constant."[8]

There seems to be evidence that these decay rates are different now from what they were in the past. Dr. John D. Morris wrote that "there are several clues that past rates have changed, or that some other process dominated. For example, the existence of short half-life polonium halos in rock have been used by many to argue for rapid formation (i.e. creation) of host rocks. Even evolutionists admit that the halos are a mystery."[9] According to Huse, "polonium 218 has been considered a daughter element of the natural decay of uranium but through the works of Dr. Gentry and other researchers, polonium halos have been found in mica and fluorite without any evidence of parents. In other words, it was primordial-present in the original granite from the very beginning. Also, and most significantly, polonium halos should not exist at all because of their extremely short half-lives. Polonium 218 has a half-life of only 3 minutes. If the evolutionist's interpretation was correct and the rock formations gradually cooled over millions of years, the polonium would have decayed into other elements long ago."[10]

A third assumption involved in radioisotope dating is that daughter and parent material are basically immobile within a rock sample. However, this is usually an erroneous assumption. There are actually many ways that these materials can become mobile. This happens, most often, by groundwater leaching. The Flood of Genesis would have caused an unprecedented amount of leaching, on a global scale. There is no rock that was unaffected by the leaching of the Flood.[11] Sylvia Baker, who is a biologist, wrote that "uranium can be lost from the rock in other ways. Uranium is often in a form that is readily soluble in weak acid. Tests have shown that up to 90% of the total radioactive elements in some granites could be removed from the surface by leaching the rock with weak acid."[12]

Leaching is an important source of error. If uranium is leached out of a mineral sample, then the relative amounts of uranium and lead will be changed. This makes the age of the sample far too high. Uranium is found, in most igneous rocks, in a state that is quite soluble in weak acids. Much of the uranium in igneous rocks has been found to be easily removed by leaching. Furthermore, it is now recognized that most of the minerals that are radioactive did contain some lead when they were formed. So most of the early age calculations using the uranium/lead procedure are highly questionable.

Besides the unwarranted assumptions, there is another good reason to be skeptical of the results of radiometric methods. This has to do with the large experimental error that is often associated with these procedures. Radiometric methods involve a high degree of technical difficulty. According to Whitcomb and Morris, "each of the parent elements disintegrates by some process through a certain chain of elements and isotopes until it reaches a stable condition. Geochronological use of these facts requires very accurate measurements of the amounts of the various elements of the chain present in the mineral and also very accurate knowledge of the respective decay constants. The techniques for these determinations are exceedingly difficult and subject to a large error."[13]

Radiocarbon dating is another type of radiometric procedure. According to Dr. Gerald Aardsma, "radiocarbon is not used to date the age of rocks or to determine the age of the earth. Other radiometric dating methods such as potassium-argon, or rubidium-strontium are used for such purposes by those who believe that the earth is billions of years old. Radiocarbon is not suitable for this purpose because it is only applicable: a)on a time scale of thousands of years and b) to remains of once- living organisms (with minor exceptions, from which rocks are excluded)".

Radiocarbon dating has supposedly set the date of some peat deposits, and other organic materials, at more than 50,000 years. However, these results have been questioned. Dr. Aardsma wrote that "some organic materials do give radiocarbon ages in excess of 50,000 'radiocarbon years'. However, it is important to distinguish between 'radiocarbon years' and calendar years. These two measures of time will only be the same if all the assumptions which go into the conventional radiocarbon dating technique are valid. Comparison of ancient, historically dated artifacts (from Egypt, for example) with their radiocarbon dates has revealed that radiocarbon years and calendar years are not the same even for the last 5,000 calendar years. Since no reliable historically dated artifacts exist which are older than 5,000 years, it has not been possible to determine the relationship of radiocarbon years to calendar years for objects which yield dates of tens of thousands of radiocarbon years."[14]

Radiocarbon dating seems to provide evidence that the earth is relatively young. Morris and Morris wrote than "it is interesting to recognize that radiocarbon (that is, Carbon 14, the unstable isotope of natural Carbon 12), which has been the basis of the main argument for human antiquity, actually yields a strong argument in support of a young earth. Radiocarbon (or C-14) is formed in the atmosphere by the action of cosmic rays on the atoms of Nitrogen 14 in the air. However, the radioactive instability of this

C-14 causes it immediately to begin to decay to N-14, the stable isotopic form of carbon." The half-life of C-14 is about 5,700 years. So after about 35-40,000 years, in a given system, there would be almost no measurable C-14.

Morris and Morris go on to say that "if the cosmic rays were somehow shut off, so that no more C-14 could be generated, it would take about, say, 50,000 years before all the radiocarbon in the earth's atmosphere, biosphere, and hydrosphere would have decayed back to nitrogen. Similarly, if the cosmic radiation were then turned on again, at the same rate as before, it would take about 50,000 years to build the C-14 back up to a level at which it would be in a 'Steady-state', with as much being produced worldwide as the amount decaying."

Morris and Morris also wrote that "the worldwide 'assay' of radiocarbon is not yet at this level, indicating that its production has only been going on for much less than 50,000 years! It is not yet in a steady state, but is still building up in the world. Yet radiocarbon datings on human artifacts are invariably based on this assumption that, regardless of what the data show, the radiocarbon assay of the world must be in a steady state. The atmosphere and the earth are far older than 50,000 years, so the reasoning goes, and so that fact requires that radiocarbon must be in a steady state." Because of the fact that radiocarbon methods are based on a steady-state assumption, the resulting dates are too high. Organisms that lived in the past never achieved the C-14 levels that currently living organisms have. After the death of an organism the C-14 decays while the C-12 remains unchanged. So the proportion of C-14 to C-12 was thought to reflect the time since an organism died. However, there was a misconception concerning the original ratio of isotopes. So the dates derived from radiocarbon procedures are distorted. Since C-14 levels are actually rising, this distortion increases as the time that has elapsed since the death of the organism increases.[15] In fact, current radiocarbon levels indicate that the earth is, at most, about 10,000 years old.[16]

So we find that the standard radiometric methods are not really reliable. Even though there are serious problems with these methods, radiometric dating may be the primary reason that multitudes of people have rejected the book of Genesis as a historically accurate document. The flaws of radiometric dating must be recognized before it can be removed from its sacrosanct pedestal.

The Earth Must Be Young

There is considerable evidence that the earth is young, apart from the radiometric issue. The first evidence has to do with the earth's magnetic field. This magnetic field is actually losing strength exponentially. The half-life of this decay is 1,400 years. In other words, the earth's magnetic field was twice as strong 1,400 years ago as at the present time.[17] According to Morris and Morris, "if extrapolated back to 7000 years (that is 5 x 1400), which is the approximate age of the earth that is suggested by known human history, human population statistics and atmospheric radiocarbon buildup, the strength of the field would have been $(2)^5$, or 32 times as strong as it is today! It could hardly ever have been much greater than that, as the earth's structure itself would have disintegrated with a much stronger field.[18]

The strength of the magnetic field is measured by devices called magnetometers, which are placed at many locations on the earth's surface. In fact, the magnetic field has been measured in this way for over 140 years. Dr. Thomas Barnes calculated that 10,000 years was the upper limit for the age of the earth, based upon these measurements. The magnetic field is the product of processes that operate in the core of the earth, which are not affected by external influences. Thus, an assumption of a constant rate of decay is , most likely, valid.[19]

The influx of various chemicals into the world's ocean waters provides much evidence that the earth must be young. Calculations of the age of the earth, involving over 30 different chemicals, yield ages that are very disappointing to the evolutionists. The chemical composition of ocean water has been known for some time. The same is true of the rivers that flow into the ocean. The amount of a particular chemical in the ocean can be divided by the amount of that chemical that flows into the ocean annually. This simple arithmetic gives the total number of years required to achieve the concentration of that particular chemical. Such calculations yield ages that are far too high since they assume that the ocean originally contained just water. Consider salt, for example. It would be unwarranted to assume that the ocean originally contained no salt, because of the fact that so many organisms seem to be designed to live in salt water.

Some evolutionists object that these calculations, measuring the influx of chemicals will only yield the "residence time" of a chemical in ocean water. These evolutionists contend that such computations do not really measure the "build-up" time of a chemical from a zero base. They claim that these calculations yield only the residence time of that chemical. However, Morris and Morris wrote that "actually, the same calculation would yield both the build-up time and also the residence time once the 'build-up' had reached a 'steady-state' with the 'out-go'. For the idea of residence time to have any meaningful basis, however, it first needs to be shown that there is actually an efflux of the given chemical from the ocean (through atmospheric recycling, sea floor deposition, or other 'sinks') sufficient to achieve the postulated steady state. This, however, the evolutionists have not been able to show."

Morris and Morris went on to say that "at least one rather detailed study of one important element — uranium — was made a number of years ago, in an attempt to quantify all such sequestering agents and sinks along the way." In fact, ocean water contains over 4 billion tons of dissolved uranium This amount is equivalent to about 3,640 trillion grams. It has been determined that 2.88 billion grams of dissolved uranium are added yearly due to the influx of rivers. When you divide the total amount of dissolved uranium in ocean water, by this annual influx, you arrive at a maximum age of the ocean. This maximum age is approximately 1,250,000 years. This may seem to be a very long time, but according to the evolutionary chronology, it really only goes back to the Pleistocene epoch. [20]

Dudley J. Whitney devoted a substantial part of *The Face of the Deep* to the issue of dissolved substances in ocean water. He wrote that "geologists almost without exception seem to have done their figuring on the supposition that the ocean was composed of fresh water from the beginning, but this was a mistake. Geochemists now almost all believe that the ocean was saline from the beginning, and on that basis any great and continual increase in the amounts of any one of the soluble substances limits greatly the possible age of the ocean."

For example, the surprisingly low quantity of sulfate in the ocean indicates that the earth is of a relatively young age. Sulfate enters the ocean from the air as ammonium sulfate, $(NH_4)_2 SO_4$. It can also enter from the air as an acid or an oxide. According to Whitney, "the nitrogen of the ammonium is taken up by marine plants and the sulfur ends up as potassium or magnesium sulfate. What is taken up by plants goes back into solution when they are consumed, mere traces remain in marine muds, though great quantities would be found there if a large part of the sulfur of sulfates went out of solution permanently. Therefore we have the condition that sulfate from river flow is being added to the ocean more rapidly that any other soluble substance and, considering the enormous output of sulfur and its compounds from volcanic action, most of it entering the ocean directly and not from river flow, the ocean cannot be old, and a different history for the earth from what the geologists offer us should be believed."[21]

Mr. Whitney went on to say "note also the great quantity of sulfurous gases that are brought to the surface in volcanic action, to say nothing of the numerous sulfur springs here and there over the lands. If water has been coming from within the earth for hundreds of millions of years, sulfur must have been entering the ocean for all that time and the ocean then would contain many times as much sulfate as it does." Furthermore, the presence of sulfur in ocean water keeps a number of elements, such as chlorine, magnesium and potassium in solution. Whitney also wrote that "the great increase of sulfuric oxide as an acid in the ocean each year would tend to bring these bases into solution instead of precipitating them."

Another important indication of the earth's true age has to do with the amount of calcium carbonate, $CaCO_3$ that is removed from the land. The enormous amount of calcium carbonate that is eroded from the land's surface, yearly, actually limits the earth's maximum possible age. Concerning the amount of $CaCO_3$ that is removed each year, D.J. Whitney wrote that "it is the equivalent of about three-fourths of a ton of limestone per year for each inhabitant of the globe. At this rate of denudation a few thousand years would remove most of carbonate available to easy weathering, and after that the calcium removed by the weathering of granite, basalt, and such massive limestone as was exposed to the air would be comparatively small. The great amount of calcium carbonate now carried from the lands to the sea by river flow can only be accounted for by the recent uplift of the lands."[22]

The study of population growth provides more evidence that the earth must be relatively young. According to Dr. John D. Morris, "observation of earth's population and population growth likewise supports the young earth. Given the total number of people on earth today, now approaching 6 billion, and its present rate of population growth of about 2% per year, it would take only about 1,100 years to reach the present population from an original pair, which is of the same order of magnitude as the time since Noah's Flood – at least it's within the right ballpark."

Evolutionists claim that humans appeared hundreds of thousands of years ago, perhaps even a million years ago. If mankind has existed for one million years, there should now be approximately 10^{8600} people living. This would be true if present rates of growth are typical. This is, obviously, an absurdly large number. However, the calculation of such an enormous number is based on uniformitarism. This is an excellent example of what can happen if uniformitarian assumptions are taken seriously. You can start from an original pair that supposedly existed one million years ago. Then, when you calculate the population growth that is necessary to reach 6 billion, you arrive at an annual growth rate of about 0.002%. Such a growth rate is very different from what has been measured during recorded history. When you consider these calculations, it becomes very difficult to believe that mankind has existed for a million years, or even half of a million years.

Furthermore, this growth rate over a million years would mean that a truly astronomical number of humans would have lived and died. Dr. John Morris also wrote that "the number is so large, it is meaningless, and it's approximately the number which could just fit inside the volume of this entire earth! If all these people lived and died, where are their bones? Why are human bones so scarce?" The study of population growth, per se, does not provide conclusive proof of a relatively young earth. However, these population calculations do seem to be very compatible with a literal interpretation of the Bible.[23]

The study of meteoritic dust provides more evidence that the earth is relatively young. A scientist at the Swedish Oceanographic Institute, named Hans Peterson, found that over 14 million tons of meteoritic dust settles to earth each year. If the earth is really billions of years old, there should be a deep layer of this dust on the surface. This layer would be dozens of feet thick. Obviously, no such layer exists. However, some evolutionists have argued that this layer of meteoritic dust is absent due to processes of crustal mixing.[24]

Even if crustal mixing processes could incorporate all that dust into the earth's crust, there is another big problem concerning the composition of meteoritic dust. According to Huse, "Geologists might be tempted to argue that erosional and mixing processes can account for the absence of the meteoritic dust layer. However, this explanation is unsatisfactory and easily refuted as the composition of the meteoritic dust is very distinctive, particularly in its content of nickel and iron. For example, nickel is a very rare element in the earth's crust and even more scarce in the ocean. But the average nickel content of meteoritic dust is approximately 300 times as great as in the earth's crust!" Although the influx of meteoritic dust is probably subject to a wide fluctuation, according to uniformitarism theory there should have been a truly stupendous amount of this dust deposited on the earth by now. However, the nickel content of the crust does not reflect that this has taken place. The small amount of nickel that has been found in the crust is very compatible with a young-earth model.[25]

The sedimentary rocks also provide evidence to support the young-earth model. Dr. Henry Morris wrote that "these give evidence of rapid deposition and so provide prima facie evidence of flood flows, not quiet deposition in stationary bodies of water. As we shall see, these great sedimentation beds also show strong evidence of essentially contemporaneous, continuous deposition rather than intermittent deposition separated by long ages of quiescence." Apart from evolutionary assumptions, there is no good reason to believe that the world's sedimentary rocks were deposited over long ages of geologic history. Dr. Morris went on to say that "there is no reason whatever why rapid (or catastrophic) formation of these beds would not provide as satisfactory an explanation, and as fully in accord with the assumption of uniform natural law, as would slow deposition over millions of years."[26]

The lack of worldwide unconformities is an indication that the rock strata were deposited continuously. Unconformities represent gaps in the deposition of sediments. Typically, an unconformity is formed when an uplift causes sediments to rise out of the water so that deposition ceases and erosion begins. According to uniformitarian theory, these unconformities should be worldwide in extent. Unconformities should separate the supposed "ages" of geologic history everywhere. However, these worldwide unconformities are not actually observed except at the bottom of the real-world geologic column. Therefore, there weren't any time gaps in the deposition of sediments that formed the rock strata, except in fairly local areas.[27]

Dr. Larry Vardiman wrote that "as the continents rose and the sea floors sank at the end of the Flood to form a new isostatic adjustment of the earth's crust, some of the unconsolidated sediments were eroded off the continents into the oceans and settled on the floor close to the continental boundaries. The continental sediments solidified into rock (lithification) forming the sedimentary rocks observed all over the earth today. On the ocean floor the deeper sediments lithified under the influence of higher pressures and temperatures which removed or chemically altered the water present in the sediments."[28]

The study of these sea-floor sediments gives more proof that the earth isn't nearly old enough for evolution to have taken place. 27.5 billion tons of sediments enter the ocean each year. About 410 million billion tons of sediments can be found on the bottom of the ocean. An age of 15 million years can be calculated by simple division. If no sediments were in the ocean to begin with, and if the rate of sedimentation was constant, it seems fair to say that 15 million years would be the approximate age of the basins of the ocean. If the oceans are the age that the uniformitarian geologists claim they are, these basins would already be full of sediments.[29]

Most of the sea-floor sediments were probably deposited during the Flood. A global Flood, such as that described in the seventh and eighth chapters of Genesis, would have produced enormous quantities of sediments. These sediments would have been deposited in the oceans and also on the continents. However, a much smaller amount of sediments were evidently produced during the creation week as the continents and the oceans were separated (Genesis 1:9, 10). But most of the sediments that are now in the ocean basins have been there since the Flood.[30]

The Cretaceous/Tertiary boundary may serve as a partition between the events associated with the Flood and the events of the post-Flood era. Many evolutionists would say that this boundary represents an important change in earth history.[31] Dr. Vardiman wrote that "the young-earth Flood model suggests that warm climates and the lack of permanent ice in the polar regions occurred before the Flood. Most, if not all, of the sedimentary layers were produced by the Flood. At some level high in the geologic column, possibly at or near the end of the Cretaceous period, the Flood waters receded, accumulation of sediments decreased, and the climate cooled."[32]

The lack of soil layers is another indication that the earth is relatively young. Uniformitarian geologists claim that the sedimentary strata represent periods of deposition that were separated by long time gaps. If this is true, a complete soil profile would have had ample time to develop between these periods of deposition. If uniformitarian assumptions are correct, the world's sedimentary rocks should contain remnants of these soil layers. These soils would have been buried by the alleged successive periods of deposition, Furthermore, these ancient soils should serve as lines of demarcation between the various, alleged, periods of geologic history. However, soil layers are not present in the sedimentary strata. Some evolutionists have claimed that the underclays, which are found below coal seams, were once soils which have been leached. But these underclays are not the same thing as soils.[33] Underclays do not have a soil profile with a sequence of horizons. Any given area of soil should have a profile. According to Whitcomb and Morris "related to-the nature of the *stigmaria* has been the question of the 'underclays', which are supposed to be the fossil soils in-which the coal-swamp vegetation grew. However, recent careful studies on the chemical and physiological nature of the underclays show this to be highly improbable."[34]

The absence of soils in sedimentary rocks throws serious doubt on uniformitarian assumptions. The best explanation for this lack of soils is that those ancient soils never existed. Dr. John Morris wrote that "standard geology tells us that land surfaces supporting lush life have been here continuously for hundreds of millions of years. Where, then, are the soils? A better explanation is that only one soil existed before the depositional episode which resulted in the majority of the geologic record."[35]

Another indication of a relatively young earth has to do with petroleum and natural gas. These two fuels are contained in underground reservoirs at high pressures but, research has shown that 10,000 years is approximately the maximum amount of time that those pressures can be maintained. Geologists claim that oil was formed very slowly, beginning many millions of years ago. However, the high pressures in the world's oil wells completely eliminate this belief as a possibility.[36] According to Dr. Scott Huse,"calculations based upon the measured permeability of the cap rock reveal that oil and gas pressures could not be maintained for much longer than 10,000 years."

Uniformitarian geologists claim that it requires millions of years to form gas and oil. But modern research has shown that this is another false assumption. Dr. Huse wrote that "recent experiments have demonstrated conclusively that the conversion of marine and vegetal matter into oil and gas can be achieved in a surprisingly short time, For example, plant derived material has been converted into a good grade of petroleum in as little as 20 minutes under the proper temperature and pressure conditions."[37]

There is a very interesting aspect of the research concerning oil formation. It has been found that hot water acts as a catalyst for ionic reactions which are involved in the formation of oil. The heat produced by the volcanism associated with the Flood, along with the enormous amount of organic material that was transported and buried during the Flood, would surely have been an ideal situation for the formation of oil.[38]

Uniformitarian geologists have had difficulty explaining the fact that oil is found throughout the geologic column. This situation has made the search for new oil reserves very difficult. According to Whitcomb and Morris, "oil has been found in rocks of practically all geologic ages except the Pleistocene. It is a feature essentially common to all the stratified rocks and, therefore, cannot be easily located by means of the usual stratigraphic and paleontologic criteria for identifying rocks. This fact also gives strong testimony that such a universal phenomenon as oil found as it is in all the rock systems must have a universal explanation. The conditions of its formation must have been essentially the same everywhere. Rather than supporting thereby the concept of uniformity in time, this fact seems rather to evidence the fact of uniformity of manner of origin and formation and thereby to imply one global event which somehow brought about the genesis of all the great oil reservoirs of the earth's crust!"[39]

Likewise, scientific research has shown that uniformitarian theory does not adequately explain coal formation. It does not actually take millions of years of heat and pressure to form coal. Research has shown that coal and coal-like substances can be produced in the laboratory, under certain conditions. The main requirement is heat. Essentially, the process involves heating organic material while isolating it from oxygen. In this way, the organic material will not ignite. Heat is necessary to get this process started but, once started, it can provide its own heat. It has been found that water, which is very hot, can provide the necessary heat.[40]

The peat-bog theory is in serious trouble. Dr. Morris wrote that "there is no actual evidence that peat is now being transformed into coal anywhere in the world. No locality is known where the peat bed, in its lower reaches, grades into a typical coal bed. All known coal beds, therefore, seem to have been formed in the past and are not continuing to be formed in the present, as the principal of uniformity could reasonably be expected to imply."

In fact, the evidence actually shows that the enormous accumulations of plant material were deposited by water. Seams of coal have almost always been discovered in stratified deposits. According to Morris, "coal is the end product of the metamorphism of tremendous quantities of plant remains under the action of temperature, pressure, and time. Coal has been found throughout the geologic column and in all parts of the world, even in Antarctica. Many coal fields contain great numbers of coal-bearing strata, interbedded with strata of other materials, each coal seam having a thickness which may vary from a few inches to several feet."[41]

Considering the alleged ice ages with regard to the age of the earth, there is evidence that only one ice age took place. This ice age corresponds to that of the Pleistocene epoch, on the evolutionary time scale. The Pleistocene came just before the Recent epoch, which is considered to be the time of recorded history. The evolutionists claim that this ice age began roughly two million years ago and that 11,000 years ago it came to an end. However, most creationists believe that the Ice Age started shortly after the Flood and lasted for less than 1,000 years.

There is no convincing evidence that the alleged earlier ice ages took place. Evolutionists claim that varved sediments and tillites in Precambrian and Permian strata show that ice ages occurred during these periods of geologic history. However, this is just an interpretation and is not conclusive proof. Ham, Snelling and Wieland wrote that "varves, for instance, are thin, rhythmic silt and clay layers that are usually thought to represent slow sedimentary processes of deposition within a glacial lake below a glacier. Each couplet of layers is considered to represent annual repetitions under summer and winter transport of sediments into the lake. However, Lambert and Hsu have presented evidence from a Swiss Lake that these varve-layers form rapidly by catastrophic, turbid water underflows. At one location, five couplets of these varves formed during a single year. At Mount St. Helens in the USA, a 25-foot (7.6 meter) thick stratified deposit consisting of many thin laminae akin to varves was formed in less than one day (June 12, 1980)."[42]

A till has been defined as "an unstratified deposit of gravel, sand, and clay which is considered evidence of glacial origin."[43] A tillite is a till that has hardened into rock. It amounts to a somewhat chaotic aggregation of sand and gravel, with some larger rocks. A tillite has a matrix of clay. the problem with interpreting a tillite as being evidence of a glacial period is that there are other deposits, which are not of glacial origin, that closely resemble a tillite. These other deposits also exhibit a lack of stratification and a lack of sorting according to size. It is very easy to confuse tills with other, very similar deposits.[44]

Furthermore, tills may have been deposited by moving water rather than ice. According to Whitcomb and Morris, "many of the evidences for ice sheets such as tills, striations, etc., can be interpreted as well or better in terms of catastrophic diluvial action. This could easily be true of other supposed glacial features such as kames, eskers, erratic boulders, etc., as well. Glacial geologists have never answered the cogent criticisms of Sir Henry Howarth, President of the Archaeological Institute of Great Britain near the close of the nineteenth century, who amassed a tremendous amount of evidence that most of the supposed ice-sheet deposits may have been formed by a great flood sweeping down from the north."[45]

The existence of alpine glaciers, ice caps, and various glacial land forms indicate that the ice age must have taken place relatively recently. Dr. Velikovsky believed that the ice actually began to retreat during historical times.[46] Dr. Whitcomb wrote that "the end of the ice age is much more recent than was once speculated, and there is much evidence now available that there was only one great glaciation, not four."[47]

To some extent, the earth is still in an ice age. Immense sheets of ice are located in the higher latitudes near the North and South Poles. If all the ice in the Antarctic Ice Sheet melted, sea-level would rise by over 50 meters. This enormous mass of ice reaches a maximum altitude of about 4000 meters, near its center. The Antarctic Ice Sheet has a very strong influence on the weather of the Southern Hemisphere. There is another huge mass of ice in the North Hemisphere. This is the Greenland Ice Sheet, which has its center at about 75^0 north latitude. There is no permanent ice sheet at the North Pole because there is no large land mass there. The Greenland Ice Sheet, which covers 80% of Greenland, is only surpassed in size by the Antarctic ice mass. The Greenland Ice sheet covers over 1.7 million square kilometers. The surface of this ice mass has an average altitude of 2,135 meters.

The Greenland Ice Sheet is an important issue in the creation vs. evolution controversy. This issue mainly has to do with the interpretation of ice cores that have been taken out of the ice sheet. Ice core drilling was begun in 1956 to study the polar ice sheets. Evolutionists claim that the Greenland Ice Sheet is the result of about 160,000 years of accumulation. However, creationists believe that this mass of ice began to form soon after the Flood.[48] According to Dr. Larry Vardiman, who is chairman of the Astrogeophysics department at the Institute for Creation Research, "a literal Biblical chronology of earth history would require the 'Ice Age' to follow the great Flood described in Genesis 6-9 because the Flood produced the sedimentary-rock strata evident all over the earth, and the 'Ice Age' evidence lies on top of these strata. A Biblical chronology like that of Ussher (1786) would place the 'Ice Age' between 1000 and 3000 years BC. Other strict Biblical chronologies might move the 'Ice Age' back in time by a few thousand years, but none would project the 'Ice Age' back to 18,000 BP. A literal interpretation of the Bible could not in any way accommodate 'Ice Ages" extending 160,000 years in the past."[49]

Dr. Vardiman wrote in chapter one of *Ice Cores and the Age of the Earth* that "direct evidence exists in the ice for annual layers which can be traced back several thousands of years. However, due to properties of the ice sheets (accumulation rates, firnification, and thinning), the identification of annual layers is limited to less than 10,000 years BP. Firnification is the formation of compacted snow by melting, re-freezing, and sublimation, in which characteristics of a snow layer are sealed off from effects above. Thinning is the gradual squeezing of ice layers as the accumulating weight of ice above forces ice below to move laterally. Below the level at which annual layering can be directly observed, models have been constructed to extend the chronology as far back as possible." The level, below which layering is not directly observable, is about 1100 meters below the surface of the ice. This would be roughly halfway down through the ice sheet. BP refers to the total number of years before present.

Dr. Vardiman goes on to say in that same chapter that "it is in constructing models to extend the geochronology that the preconceived view of earth history enters the interpretations. If one does not know the meteorological conditions in the past, and knowledge of these conditions is necessary to extend the ice core record back in time, it is common practice to assume that prior conditions were similar to those of today. The geological community uses a phrase for this uniformitarian assumption 'the present is the key to the past.'" In other words, uniformitarian geologists based their estimate of the age of the great polar glaciers on the assumption of a very gradual, uniform rate of accumulation.

The question is whether or not this is a warranted assumption. According to Vardiman, "if the meteorological conditions in the past were considerably different than they are today, such a model would not adequately represent the real chronology. For example, if accumulation rates near the bottom of the core were much greater than today, then the great ages found by the previous method would be invalid and much older than the actual fact." The Flood actually caused a tremendous amount of precipitation, and therefore massive accumulation in a relatively short period of time after the Flood. Dr. Vardiman also wrote that "as the polar regions cooled following the flood, approaching the temperature distribution observed today, large quantities of water vapor would have been evaporated from the warm oceans and deposited over cold polar regions. Precipitation rates would have been enormous initially, slowing to that observed today."[50]

Mr. Whitney also believed that glacier formation followed the cooling of ocean water near the poles. He wrote that "not until the polar seas became cold and ice formed over parts of the ocean would a quick growth of glaciers begin there. Then there would be a fairly rapid fall in sea level until the accumulation and loss of ice in these areas reached an approximate equilibrium. The situation therefore seems not only to discredit conclusively the ice-age theory in favor of Deluge geology, but it seems to indicate that Ussher's estimate for the date of the Deluge is approximately correct."[51]

In addition to these indications that the earth is young, there is much evidence that the solar system and universe are relatively young. Let's consider the comets first. Comets are in highly elliptical orbits around the Sun. According to Carl Sagan, "a comet is made mostly of ice-water (H_2O) ice, with a little methane (CH_4) ice, and some ammonia (NH_3) ice." A comet also has a kind of a nucleus which is not composed of ice.[52] Dr. Sagan also wrote that "somewhere between the orbits of Jupiter and Mars it would begin heating and evaporating. Matter blown outwards from the Sun's atmosphere, the solar wind, carries fragments of dust and ice back behind the comet, making an incipient tail." A comet's tail points away from the Sun, coming and going.

Carl Sagan wrote that "a fairly typical comet would look like a giant tumbling snowball about one kilometer across."[53] However, Halley's Comet has been found to be much larger than that. A spacecraft actually photographed this famous comet in 1986 from a distance of about 600 kilometers. The solid part of Halley's Comet was found to be about 16 kilometers long and was shaped somewhat like a peanut. Its nucleus was dark and granulated.[54] Each time Halley's Comet passes the sun, it loses part of its mass. This is true of all comets. According to Huse, "each time a comet orbits the sun, a small part of its mass is 'boiled off'. Careful studies indicate that the effect of this dissolution process on short-term comets would have totally dissipated them in about 10,000 years. Based on the fact that there are still numerous comets orbiting the sun with no source of new comets known to exist, we can deduce that our solar system can not be much older than 10,000 years."[55]

Since the comets, as well as the planets, are in orbits around the Sun, it just seems reasonable that all of these heavenly bodies are of about the same age. According to Whitcomb and Morris, "since comets are very definitely a part of the solar system, the natural inference would be that the maximum age of the comets would also be the maximum age of the solar system." Whitcomb and Morris went on to say that "evidently, all the known comets can be expected to break up and vanish within a time which is geologically very short."[56] Meteors are believed to be the products of the disintegration of comets.[57]

The disintegration of the comets presents a big problem to evolutionary theory. That's why evolutionists claim that there is a vast cloud of swarming comets far beyond the orbit of Pluto. They call this alleged source of new comets the Oort Cloud. Morris and Morris wrote that "they have invented an imaginary 'Oort Cloud' of 'wannabe' comets somewhere on the distant edges of the solar system. Every now and then, the gravitational pull of some nearby star, or of something else, supposedly pulls one of these comets-in-waiting out of the cloud, and sends it spinning into orbit around the sun. When that comet eventually disintegrates and falls into the sun, another will be fortuitously released from the cloud." The main problem with the Oort Cloud theory is that it is just purely theory. The fact is that no astronomer, looking through any telescope, has ever actually seen this alleged source of new comets.[58]

There is evidence that the sun, which is an average of about 93,000,000 miles from Earth, is not nearly as old as was once thought. Keith Davies, who is a Christian educator in Canada, wrote that "when a star like our sun is very young, its enormous output of energy is provided by gravitational contraction. As it grows older, the models show that the source of energy should change over to that of nuclear fusion as it slowly develops a very hot and dense core." He also said that "the standard model of the sun assumes that it is around 5 billion years old and that it has already passed into its nuclear burning stage. This makes it all the more extraordinary that in 1976 a team of Russian astronomers showed how their research pointed clearly to the startling fact that the sun does not even seem to possess a large dense nuclear burning core. Instead, their results showed the sun as bearing the characteristics of a very young homogenous star that corresponds with the early stages of the computer models."

This article appeared in volume 259 of *Nature* and was written by Severny, Kotov, and Tsap. These scientists concluded that energy is not generated in the sun by nuclear reactions. This conclusion was based on their research concerning the sun's oscillations. They found the global oscillation period to be much larger than what would be expected if the sun had a large, dense core. They were surprised to see that the global oscillation period was actually 2.65 hours. According to Mr. Davies, "The predicted oscillation of 2 hours 47 minutes is particularly important as being a key distinguishing feature of a young homogeneous star. The Russian astronomers were certainly startled to find that their observations of the Sun were showing large and remarkably stable global oscillations with a period of 2 hours 40 minutes very close to that predicted for a young homogeneous sun."

The measured amounts of lithium and beryllium in the sun are more evidence that it is relatively young. Keith Davies wrote that "we know that lithium would be destroyed in around 7,500 years when the central temperature of a young star reaches 3 million degrees. However, the Sun still has its normal abundance of beryllium which is destroyed at a temperature of 4 million degrees. If the Russian scientists are correct in assuming that the Sun is homogeneous, then this means that the temperature throughout the whole Sun must be far lower than the 15 million degrees required for the Sun to be an old main-sequence star." It is highly significant that there is still a detectable amount of lithium in the Sun. If it should all be gone after about 7,500 years, then the Sun can't be that old.[59] Along these same lines, Walter Wilson said that, "those with any knowledge at all of chemistry or mineralogy will know without a doubt that metals of this character, and many others, could never withstand the heat of the sun for any length of time."[60]

Another indication that the sun is a young star is its lack of neutrino emission. A neutrino is a subatomic particle that is supposedly produced by nuclear fusion which occurs within the Sun. They are believed to be highly penetrating particles which can pass right through the Earth and emerge on the other side. Neutrinos do not interact very much with anything. However, scientists believe that neutrinos do occasionally interact with chlorine to form argon. So, theoretically, the conversion of chlorine to argon could be used to measure the Sun's output of neutrinos. A rather bizarre experiment was devised to test this theory. Carl Sagan wrote that "neutrinos can on rare occasion convert chlorine atoms into argon atoms, with the same total number of protons and neutrons. To detect the predicted solar neutrino flux, you need an immense amount of chlorine, so American physicists have poured a huge quantity of cleaning fluid into the Homestake Mine in Lead, South Dakota. The chlorine is microchemically swept for the newly produced argon. The more argon found, the more neutrinos inferred. These experiments imply that the Sun is dimmer in neutrinos than the calculations predict. There is a real and unsolved mystery here."[61] To an evolutionist's way of thinking this certainly is a mystery.

Neutrino flux from the sun probably did not have anything to do with the argon that was found in the previously mentioned experiment. It just seems sensible that the presence of argon was due to some kind of contamination, possibly from the atmosphere. Argon makes up about one percent of Earth's atmosphere. After all, the pouring of liquid chlorine into the Homestake Mine doesn't exactly qualify as being a situation of controlled laboratory conditions. It is very questionable that the Sun is producing neutrinos at all. According to Morris, "most astrophysicists believe they are derived from the energy released by thermonuclear fusion processes deep in the heart of the sun. However, there are many unanswered problems associated with this explanation, one of the most important being the fact that the solar neutrinos which would be produced by these reactions are apparently not being produced at all, according to the very sophisticated techniques that have been devised to measure them. This would mean almost certainly that the sun's heat is being generated by the gravitational collapse of its gases toward the center."[62]

There is also evidence that the moon is relatively young. It occupies an orbit which is about 237,000 miles away. Dr. Barnes wrote that "from the laws of physics one can show the moon should be receding from the earth. From the same laws one can show that the moon would have never survived a nearness to the earth of less than 11,500 miles." He also wrote that "the tidal forces of the earth on a satellite of the moon's dimensions would break up the satellite into something like the rings of Saturn. Hence the receding moon was never that close to the earth. The present speed of recession of the moon is known. If one multiplies this recession speed by the presumed evolutionary age, the moon would be much farther away from the earth than it is, even if it had started from the earth. It could not have been receding for anything like the age demanded by the doctrine of evolution." [63]

The complex ring system of Saturn seems to provide more evidence that the solar system must be relatively young. The rings of Saturn are located on the same plane as its equator. These rings have a diameter of about 160,000 miles. They appear to have a continuous surface but are actually composed of innumerable separate pieces. These pieces are a mixture of rock and ice. Much of it seems to be water-ice but there are other frozen chemicals. The pieces that are the closest to Saturn move faster than those farther out. The slight differences in speed between adjacent particles keep them from clumping together. The separate pieces, in the rings of Saturn, actually form an extremely vast swirl around that planet. It is difficult to imagine how such a delicate arrangement could have remained intact for billions of years. Dr. Donald DeYoung wrote that "there are actually hundreds of narrow individual rings, perhaps kept in place by small 'shepherd moons' between them. Astronomers find it incredible that such intricate detail has remained in place for billions of years, although an evolutionary view of long ages leaves little choice."[64]

The existence of groups of stars that are gravitationally bound is evidence that the universe must be relatively young. Dr. Huse wrote, "Star clusters serve to indicate a young age for the universe. A star cluster contains hundreds or thousands of stars moving, as one author has put it, 'like a swarm of bees'. They are held together by gravity, but in some star clusters, the stars are moving so fast that they could not have held together for millions or billions of years. Thus, the presence of star clusters in the universe indicates that the age of the universe is numbered in the thousands of years."[65]

Speaking of stars, evolutionists contend that since the earth receives light from very distant stars, the universe must be extremely old. The earth receives starlight from stars that are many millions of light-years away. A light-year is actually a measure of distance. It is the distance that light can travel in one year. However, certain assumptions are involved in calculating the age of the universe from starlight. According to Ham, Snelling, and Wieland, "we do not actually observe an old universe we observe certain facts, but from these we can deduce an age only if we make certain assumptions (starting beliefs). Even though they may seem reasonable, they are nevertheless beliefs, so we should look carefully at each of these assumptions in turn, remembering that great humility is in order. We do not have access to infinite knowledge; only God Himself does. The history of science shows that what is deemed certain in one generation may be totally overthrown by new information in the next."

The first assumption that is involved, in deducing an immense age of the universe from starlight, is that the enormous distances of stars have been stable. However, distant stars may have been much closer in the past. In fact, the Bible seems to refer to an expanding universe in several places (e.g. Isaiah 42:5 and Jeremiah 10:l2). Ham, Snelling, and Wieland wrote that "there exists a solution to Einstein's gravitational field equations which may allow a very rapid expansion of space and thus all things in it. If the mathematics continues to work out, this would raise the possibility of the universe inflating to its present size in less than 6,000 years. Thus, the light began its journey when galaxies were much closer (and also much smaller and less energetic) than today. Such an expansion would involve all matter symmetrically, and would not be detectable by those 'on the inside' as it were. However, it could account for the observed redshifting in the wavelengths of this stretched starlight."[66]

The Big Bang theory is not the only explanation for an expanding universe. Dr. D. Russel Humphreys, who is a highly respected physicist, wrote that "after many years of studying the evidence, I am convinced that the observations indicate that the universe has indeed expanded significantly, by a factor of at least one thousand. There also appears to be Scriptural evidence for such an expansion." For example, the Bible says in Isaiah 4O:22 that "It, is he that sitteth upon the circle of the earth, and the inhabitants thereof are as grasshoppers; that stretcheth out the heavens as a curtain, and spreadeth them out as a tent to dwell in."[67] Expansion of the universe may be God's way of providing stability. According to DeYoung, "expansion of the universe is only one possible explanation for the redshift of starlight. If this theory is correct, God surely has his own purposes for creating a universe in outward motion. Perhaps it provides stability: in a static universe, gravity would eventually cause all galaxies to collapse inward."[68]

A second assumption that is involved in determining the age of the universe from starlight is that it is completely understood how light travels over great distances. However, it has been proposed that light travels in a curved pathway. This curved pathway exists in a mathematical framework that is called Riemannian space. Along such pathways, it is believed that distant starlight could reach the earth in less than 20 years. Riemannian geometry deals with curved space-time. Dr. Morris wrote that "one possibility is that light travels in a type of curved space called Riemannian space even though geometric space is flat. Calculations for this type of geometry have indicated that light coming from an infinitely distant source would reach the earth in less than sixteen years. These conclusions resulted from a study almost thirty years ago by two very competent evolutionary astrophysicists and electrodynamacists, P. Moon and D.F. Spencer, associated at the time with the Massachusetts Institute of Technology. However, most astronomers have rejected this idea because of its compatibility with a young universe."[69]

Dr. Humphreys has said that the problem of light from distant stars can be explained by Einstein's relativity. Dr. Humphreys wrote that "general relativity (GR) has been well established experimentally, and is the physics framework for all modern cosmologies. According to GR, gravity affects time. Clocks at a low altitude should tick more slowly than clocks at a high altitude and observations confirm this effect, which some call gravitational time dilation. (Not to be confused with the better known velocity time dilation in Einstein's special relativity theory.) For example, an atomic clock at the Royal Observatory in Greenwich, England, ticks five microseconds per year slower than an identical clock at the National Bureau of Standards in Boulder, Colorado, both clocks being accurate to about one microsecond per year. The difference is exactly what general relativity predicts for the one-mile difference in altitude."

This means that the measurement of time depends on the frame of reference. Dr. Humphreys also wrote that "what this new cosmology shows is that gravitational time distortion in the early universe would have meant that while a few days were passing on earth, billions of years would have been available for light to travel to earth. It still means that God made the heavens and the earth (i.e. the whole universe) in six ordinary days, only a few thousand years ago. But with the reality revealed by GR, we now know that we have to ask six days as measured by which clock? In which frame of reference?" [70]

A third assumption is that the speed of light has always been the same as it is now, about 186,000 mps. It is impossible to prove this assumption. Obviously, no scientist was there to measure the speed of light six thousand years ago. The slowing down of light may help to explain the redshift of starlight and the background radiation in space. This was proposed by a Russian scientist named Troitskii. He believes that the speed of light may have been nearly infinite at one time.

The fourth assumption that is involved, when evolutionists say that starlight proves an extremely old universe, is that the universe was not created with the appearance of age. However, this is more of a presumption than an assumption. According to Ham, Snelling, and Wieland, "this assumes that God is not a God of infinite power and so is the most presumptuous of all. Biblical creation by definition is a miracle using processes which are not now in operation. God is not dependent on the physical laws which we observe in the present, since He instituted these Himself, to operate with their general regularity (which also exhibits His sustaining power) only after the end of the six-day period."[71]

The appearance of age or history is actually inherent in creation. There could not be any form of creation without having, from the first, at least some appearance of age. It is this initial appearance of age that is often interpreted to be the result of evolution. Furthermore, some of the miracles of Christ involved the appearance of age. For example, in Matthew 14:14-

21 there is an account of the miraculous feeding of thousands of people. In this passage, many thousands of loaves and fishes were created. This food had an appearance of age. The barley loaves consisted of grains that were not harvested from a field. The fish, that were served, had neither been hatched from an egg nor caught in a net.

In conclusion, there is actually much evidence that the Earth must be relatively young. The study of Earth's magnetic-field shows that 10,000 years is the maximum possible age of the Earth. Likewise, the build-up of radiocarbon in the atmosphere and biosphere indicates that 10,000 years is the maximum age of the Earth. The study of underground oil reserves shows that about 10,000 years is the maximum amount of time that oil can stay in those reservoirs under high pressures. The observation of the growth of human populations is consistent with a young Earth. The measurement of the chemicals entering the ocean gives ages that are too short for Darwinism. The total absence of ancient soils, in rock strata, throws serious doubt on uniformitarian assumptions. The disintegration of the comets gives a maximum possible age of the solar system of about 10,000 years. The fact that the sun still contains a detectable amount of lithium seems to limit the age of the sun to about 7,000 years. The gravitationally bound groups of stars indicate that the universe cannot be of a great age Furthermore, the polonium halos seem to show that creation was virtually instantaneous. The only adequate explanation for the existence of these small colored spheres that appear in some rocks is that those same rocks just suddenly appeared, out of nothing!

Saturn

CHAPTER 11

CREO EX NIHILO

Creo Ex Nihilo means to create out of nothing, in Latin. *Ex nihilo* (out of nothing), in reference to creation, was probably first used by Irenaeus in the second century.[1] Irenaeus was a Christian leader who was born in Asia Minor. His parents were Greek. He is believed to have been a student of Polycarp, bishop of Smyrna.[2] Irenaeus helped to organize a canon of Scripture. This defender of the Faith wrote a book called *Against Heresies* in about AD 180, which was largely a refutation of Gnostic beliefs. Gnosticism was basically evolutionary pantheism which had a hierarchy of demonic gods. This sect relied on mysticism to gain religious insights and its devotees observed rigid asceticism.[3] Gnostics believed the world came from evil matter and that the world could not have been created by a good God. Because of the fact that Gnosticism was a type of evolutionary philosophy, Irenaeus must have been one of the first creationists of the New Testament era. There are three different strategies for incorporating modern evolutionary philosophy into Biblical interpretation. These three strategies are the Day-age Theory, the Gap Theory, and the Framework Hypothesis.

The Bible says in Exodus 20:11 that "For in six days the Lord made heaven and earth, the sea, and all that in them is, and rested the seventh day: wherefore the Lord blessed the Sabbath day, and hallowed it." Referring to the first chapter of Genesis, Rice wrote, "In brief definite language God tells what happened day by day in six days of creation. The Bible never hints that it is a 'poem' or that the language is figurative or allegorical." Some Christians advocate a creation that was progressive. By this they mean that there were long ages of time, and that God did a part of creation near the start of each age. By this means, they attempt to reconcile the first chapters of Genesis with the long ages of the geologic time scale. Advocates of the Day-age Theory claim that each day in the first chapter of Genesis represents millions of years. There are actually a number of problems with this "theory."

We find in Genesis chapter one that plants were created on the third day but that God did not create the animals until the fifth and sixth days. Now, it is a well known fact that plants and animals exist in a balance. Plants absorb CO_2 and then produce oxygen which is breathed in by animals. This oxygen is used in the body for respiration and eventually results in the release of carbon dioxide (CO_2). So how could plant life exist for millions of years in an environment where animals weren't continuously producing carbon dioxide?

Secondly, the repeated use of the phrase "the evening and the morning" in Genesis I must refer to literal days. According to Rice, "when we say a day we mean a 24-hour period generally, in which the earth revolves, with one period of light and period of darkness. Or we mean the period of light itself, the light part of the 24-hour day." God's creation of light is recorded in Genesis 1:3-5. There must have been a source of light for the Earth other than the Sun in the first part of the creation week. This is because the sun was not created until the fourth day (Gen. 1: 16). But regardless of the source of light, the rotation of the Earth would result in 24 hour periods of alternating light and dark. The six times that the phrase "and the evening and the morning" is used clearly refers to six literal days.

Likewise, there is a reference in Matthew 12:40 about Jonah being in the belly of a whale for 'three days and three nights." It would be rather silly to say that this referred to three long ages of time. This verse clearly refers to three literal days. The Bible says in Exodus 12:15 that "seven days shall ye eat unleavened bread: even the first day ye shall put away leaven out of your houses." No reasonable person would suggest that this verse refers to seven long ages of time. To try to say that the six days of creation represent six long ages of time is a strained interpretation that is not consistent with the normal use of language.

Another inconsistency with the Day-age Theory is that it says that the creation of plants and the creation of birds and other flying creatures was separated by an enormous span of years. Rice wrote that "many, many plants...can never bear fruit except they are pollinated by insects. How could they live a whole age (if the days of creation were ages of many thousands of years) before any insects were created or any birds or honeybees or other insects?"[4]

Perhaps the best evidence for the meaning of the word "day" is in Genesis 2:1-3. In that passage there are two references to the fact that God "rested" at the end of the creation week. Then over in Exodus 20:10, 11 we read that "the seventh day is the Sabbath of the Lord thy God: in it thou shalt not do any work, thou, nor thy son, nor thy daughter, thy manservant, nor they maidservant, nor thy cattle, nor thy stranger that is within thy gates: For in six days the Lord made heaven and earth, the sea, and all that in them is, and rested the seventh day: wherefore, the Lord blessed the Sabbath day, and hallowed it." Now, if God rested for a literal day at the end of the creation week, then the six days He was creating everything should have been literal days also. The context really demands such an interpretation. No reasonable unbiased person would say that the Lord's Sabbath (Ex. 20:10) refers to a very long span of time. To do so would be putting the Bible through unreasonable contortions just to please the evolutionists. Morris wrote that "the only proper way to interpret Genesis I is not to 'interpret' it at all. That is, we accept the fact that it was meant to say exactly what it says. The 'days' are literal days and the events described happened just the way described."[5]

There has been another attempt to harmonize the Lyellian view of earth history with the Bible. This second strategy is called the Gap Theory. As the name suggests, this theory says that there was an enormous span of time between Genesis 1:1 and 1:3. John C. Whitcomb wrote that, "The supposedly vast ages of the geologic timetable are thought to have occurred during this interval, so that the fossil plants and animals which are found in the crust of the earth today are relics of the originally perfect world which was supposedly destroyed before the six literal days of creation (or, rather, recreation)."[6]

The Gap Theory is also referred to as the Ruin-reconstruction Theory. It is clearly an attempt to accommodate uniformitarianism. This theory became popular in the early part of the nineteenth century. Most advocates of the Gap Theory believe that the originally created earth was destroyed by God as a consequence of Satan's fall. This destruction was by means of a global flood. This is supposedly how the world became, "without form and void" in Genesis 1:2. Morris and Morris wrote that "the whole system of modern geology has been built upon the dogma of uniformitarianism not catastrophism And it is the resulting system of geological ages that the gap theory attempts to pigeonhole between Genesis 1:1 and 1:2."[7]

Dr. Rice wrote that "the theory (1) of original creation in dateless past, (2) millions of years in which scientists can imagine, to provide the fossils, earth strata and coal beds, (3) and then a recent restorative creation of six days on the earth formerly ruined by divine judgment, is not taught in the Bible." The only way anyone could interpret Genesis 1:1-3 to mean such a thing was if they approached the study of Genesis with a strong bias in favor of Lyellian/Darwinist doctrine.[8]

Despite much effort to prove the Gap Theory, the Bible simply does not teach that there is a very long gap of time after Genesis 1:1. But let's look at some arguments for the Gap Theory anyway. One is that where the Bible says "And the earth was without form and void;" in Genesis 1:2 should have been translated "the earth became without form, and void". Dr. Whitcomb wrote in chapter five of *The Early Earth* that "while the verb hay^eta can often be translated 'became' the word order and sentence structure in Genesis 1:2 (and in a number of other passages) does not permit this translation. If it had to be translated 'became', then we would have to say that Adam and Eve 'became' naked (Genesis 2:25) and that the serpent 'became' more subtle than any beast of the field Gen. 3:1)!" He goes on to say in that same chapter that "the Septuagint translators rendered the verb 'was' and not 'became.'"[9]

Twenty prominent scholars of Hebrew were polled as to whether there was any Biblical evidence for a gap of time after Genesis 1:1. Their answer was a definite no. The phrase, "without form and void" refers to the fact that the world was simply not finished.[10] Ken Ham said that large numbers of people "really believe scientists have proven evolution and every related issue. For many people, a belief in such positions as theistic evolution, the Gap Theory and progressive creation came out of sheer pressure from their belief that scientists had proved many, if not all, aspects of evolution."[11]

C.I. Scofield suggested that Jeremiah 4:23-26 referred to the supposed great cataclysm that left the earth "without form and void." Now, the Bible does say in Jeremiah 4:23 that "I beheld the earth, and lo, it was without form, and void, and the heavens, and they had no light." However, an examination of the context shows clearly that this passage refers to the destruction of Judah and Jerusalem at the beginning of the Babylonian captivity. In Jeremiah 4:16 we read, "Make ye mention to the nations; behold, publish against Jerusalem that watchers come from a far country, and give out their voice against the cities of Judah." Then over in Jeremiah 4:29 we find that "the whole city shall flee for the noise of the horsemen and bowmen; they shall go into the thickets, and climb up upon the rocks." This verse refers to an invading army just prior to a period of exile.

Scofield also wrote that Isaiah 24:1 referred to a cataclysmic, prehistoric judgment of the earth. In this verse the Bible says that "Behold, the Lord maketh the earth empty, and maketh it waste, and turneth it upside down, and scattereth abroad the inhabitants thereof." But as with Jeremiah 4:23, an examination of the context shows that this verse refers to the destruction that resulted from warfare. God's Word says in Isaiah 24:5 that "the earth also is defiled under the inhabitants thereof- because they have transgressed the laws, changed the ordinance, broken the everlasting covenant." The Hebrew word for earth, `erets, can also be translated as "land." This fact, plus the context, indicates that this is a judgment of God on a people because of their sins.[12]

Another argument in support for the Gap Theory says that the word "made" in Exodus 20:11 really means a refashioning of the earth after the great cataclysm that supposedly happened in Genesis 1:2. According to this argument, this Hebrew verb can also mean made to appear. This word "made" in Exodus 20:11 is translated from the Hebrew word *asah*. This argument mostly has to do with whether or not there is a distinction between *asah* and another Hebrew verb *bara*.

The word "created" in Genesis 1:21 is translated from this verb bara. The Bible says in this verse that "God created great whales, and every

living creature that moveth, which the waters brought forth abundantly.".
But in Genesis 1:25 the Bible says that "God made (*asah*) the beast of the
earth after his kind." Dr. Whitcomb wrote that "surely we are not to think
that sea creatures were directly 'created' on the fifth day but land animals
were merely 'appointed' or 'made to appear' on the sixth day! All those
who hold that bara and asah cannot be used of the same kind of divine
activity have a serious problem here." In the context of the first chapter of
Genesis, the Hebrew verb asah is a synonym for *bara*.[13]

There is another argument that is used to support the Gap Theory. This
argument has to do with the meaning of the phrase "without form and void"
in Genesis 1:2. Advocates of this argument say that an omnipotent and
omniscient God would not have made the earth in a chaotic state. So the
question is whether or not "without form and void" really refers to a chaotic condition. "Without form" is translated from the Hebrew word *tohu*.
"Void" is translated from the Hebrew word *bohu*.

Dr. Whitcomb wrote, "In Job 26:7 we read that God 'stretches out the
north over empty space (*tohu*) and hangs the earth on nothing.' Certainly
we are not to find in this verse any suggestion that outer space is basically
evil! In some passages the word refers to the wilderness or desert, which is
conspicuous for the absence of life. (Deut. 32: 10, Job 6:18, 12:24, Ps.
107:40)." These two Hebrew words also appear in Isaiah 34:11. From an
examination of these verses it is reasonable to conclude that "without form
and void" in Genesis 1:2 refers to a very early stage of the creation week.
At this early stage, the earth was simply incomplete and uninhabited. There
is no good reason to believe that Genesis 1:2 refers to a state of chaos.[14]

The third strategy for the incorporation of evolutionary philosophy into
Biblical interpretation was addressed by Morris and Morris when they wrote
that "the geological ages cannot be placed before the six days of creation
(gap theory), during the six days of creation (day-age theory) or after the
six days (which, since they antedate 'man,' no one suggests at all). The
only remaining possibility is that either the six days or the geological ages
had no existence in the first place." This last possibility is what the Framework Hypothesis is all about. This hypothesis attempts to explain away the
six days of the first chapter of Genesis.

Liberal theologians almost invariably accept the geological ages as fact.
However, the validity of the supposedly sacrosanct geologic time scale has
not been proven. But because of their acceptance of the time scale, they
must reject a creation account in which God creates *ex nihilo*. Some of
these liberals say that Genesis is allegorical. Others would say that Genesis
is poetic. Some would even refer to Genesis as being supra-historical. This
hypotheses views the first eleven chapters of Genesis as a kind of rhetori-
cal framework. In this framework, issues such as the creation, man's fall,

and redemption are supposed to be symbolic of deeper meanings. This framework strategy gives Genesis some theological significance but denies that the first eleven chapters of Genesis have any historical or scientific accuracy.

One of the problems with this framework method of interpretation is that it destroys the foundation for the historical account of Abraham, which begins in chapter 12 of Genesis. The framework strategy destroys the background of Abraham who was in the lineage of the Christ because it says that the first eleven chapters are just an allegory. This method of interpretation is not consistent with a sincere belief in the Bible. Morris and Morris call the Framework Hypothesis a "method of so-called 'neo-orthodoxy', though such idealistic humanism is neither new nor orthodox."[15]

Humanism basically says that man, or actually the mind of man, is the final authority. The Framework hypothesis can be called humanism because it exalts human reason above Biblical authority. Theological liberals think that the theories of men like Charles Darwin and Charles Lyell take precedence over the Bible. That's why these liberals say that Genesis must be made to mesh with evolutionary theory. In fact, an evolutionary view of origins is the philosophical foundation of humanism. So we find that humanism and liberal theology have much in common. In the final analysis, all forms of theistic evolution are humanism.

By contrast, fundamentalists believe that the Bible is the final authority.[16] Theistic evolution means compromising Biblical authority for the sake of a humanistic concept. Fundamentalists believe that it is a serious error to tamper with the Bible. Dr. Curtis Hutson wrote that "there is an attempt on the part of scholarship, so-called, to explain from a human standpoint the language as well as the content of Scripture. For instance, you may hear a Bible teacher say, 'Paul said a certain thing because of a certain set of circumstances that surrounded him.' Now the only thing wrong with that is it is simply not true. Paul said what he said because God told him what to say."[17]

The Bible says in Deuteronomy 8:3 that "man doth not live by bread alone but by every word that proceedeth out of the mouth of the Lord doth man live." Then over in Matthew 4:4 we read, "It is written, man shall not live by bread alone, but by every word that proceedeth out of the mouth of God." These verses emphasize the importance of the whole Bible to the spiritual life of mankind. Liberal theology tries to invalidate these two verses by saying certain parts of the Bible are not really to be taken seriously. They say that it is up to man to decide what the Bible means. These liberals claim that human reason must decide which parts of the Bible are historically or scientifically accurate and which parts are not. But Dr. Rice wrote that "reason cannot tell beyond certain facts that nobody can know except by divine revelation. Who can reason and find out about the creation of the universe?"[18] Human reason cannot be totally trusted for at least two reasons. For one thing, human reason is tainted because of man's fallen and basically selfish nature. Furthermore, we are really just finite creatures, after all. Mankind will never have all the facts. Only God can see the whole picture and that's why He gave us the Bible, the actual Word of God.

According to Ken Ham, "There is one book of Moses that is referred to more often in the rest of the Bible than any other book. That book is Genesis. But in theological and Bible colleges, in Christian and non-Christian circles, which book of the Bible is the most attacked, mocked, scoffed, thrown out, allegorized and mythicized? The book of Genesis!" Theistic evolution has caused a great deal of confusion. This kind of humanism has caused large numbers of people to question the plain teachings of Scripture. If theistic evolution is valid, then how do you know what is or is not true in the Bible? Christians have not, generally speaking, realized the importance of creation as a foundational Biblical principle.

Mr. Ham also said that "If you want to destroy any building, you are guaranteed early success if you destroy the foundations. Likewise, if one wants to destroy Christianity, then destroy the foundations established in the book of Genesis. Is it any wonder that Satan is attacking Genesis more than any other book?"[19]

The Bible says in Psalm 11:3 that "If the foundations be destroyed, what can the righteous do?" It is clear that Genesis is foundational to the rest of the Bible, including the Gospel. The Gospel has been defined in I Corinthians 15:3,4 where the Bible says that "Christ died for our sins according to the scriptures; And that He was buried, and that He rose again the third day according to the scriptures." An attack on the credibility of Genesis is an attack on the credibility of the Gospel.

Dr. Henry Morris wrote, "The theological capitulation to evolution has been the forerunner and the basis of the development of modernism in religion."[20] The term modernism means the attempt by theologians to reconcile the Bible's basic teachings with what has been called contemporary thought. An attempt to undermine the supernatural aspects of the Bible is especially characteristic of modernism. Modernists deny that the Bible is infallibly inspired. They also deny that Christ was virgin born, was literally resurrected, and was actually God incarnate.[21]

The theological capitulation to which Dr. Morris referred, happened because there were religious leaders who did not consider the Bible to be the final authority. Fundamentalists do not accept human reasoning (e.g. Lyellian/Darwinist doctrine) as the final authority. Those who accept human reasoning as the final authority essentially make a god of the human mind. Unfortunately, most of the religious colleges and seminaries, in this country, teach that some form of human reasoning is the final authority.[22] The faculties of most American seminaries do not really believe that the Bible is the inspired Word of God. In other words, these professors are modernists. According to John R. Rice, "It is a part of the apostasy of this day that unconverted men, not Christians by experience, unbelieving men, not Christians in the historic Christian faith, 'once delivered', are accepted in many Christian circles. That is wrong morally and it is foolish from the viewpoint of scholarship. Those who do not revere the Bible as the Word of God are not good men and are not reliable scholars."[23]

In conclusion, evolutionary theory has had a profound theological impact. Religious leaders have attempted to incorporate evolution into Biblical interpretation. The results have been disastrous. This theological surrender to Lyellian/Darwinist doctrine has resulted in the infiltration of modernism into religious teaching. Modernism is the viewpoint that rejects the idea that Christ, the virgin-born Son of God, is an atoning sacrifice for the sins of mankind. In so doing, modernism leaves the human race without any hope at all. William Jennings Bryan said that "evolution, theistic and atheistic, carried to its logical conclusion, robs Christ of the glory of a virgin birth, of the majesty of His deity, and of the truth of His resurrection. That kind of Christ cannot save the world. We need the full stature Christ of whom the Bible tells; the Christ whose blood has colored the stream of time, the Christ whose philosophy fits into every human need, the Christ whose teachings alone can solve the problems that vex our hearts and perplex this world."[24]

CHAPTER 12

HISTORICAL IMPACT OF EVOLUTION

Charles Darwin's ideas on evolution and natural selection have had an influence on science, politics, economics, and society in general. Evolution has been used as the basis for nearly all political, economic, and social systems that grew out of a lust for power. Darwinism was probably developed as a justification for certain socio-economic practices of that time.[1] This may be the reason why the name of Charles Darwin is associated with evolution, despite the fact that the concept of evolution had been around a long time before he came along. Even the mechanism of natural selection was not originated by Charles Darwin. A number of men, including Erasmus Darwin, William Lawrence, James Pritchard, William Wells, Edward Blyth, and Denis Diderot had written about evolution and natural selection.

The political climate of that period had much to do with Charles Darwin's success. The mid-nineteenth century was a period of great societal upheaval. The French Revolution had recently ended. Karl Marx was laying the foundation of communism. Anarchism was gaining in popularity. This was also the time of the Civil War, the industrial revolution, economic imperialism, and Illuminist conspiracy. The great economic and political movements, of that time, eagerly embraced Darwinism because it could be used as a scientific justification for their humanistic philosophies. These movements were essentially anti-Christian and anti-Bible. They all wanted to undermine the credibility of the Bible in order to gain influence over the masses. They did not like the fact that so many people believed the Bible and tried to live their lives according to Biblical principles. The leaders of these movements wanted people to question the Bible, especially the doctrine of creation. If somehow people could be made to think that the creation account of Genesis had no validity, then they would soon begin to doubt the teachings of the rest of Scripture. Darwinism offered an evolutionary mechanism to explain how the world as we know it came to be.

The average person of that time had seen the changes that were brought about in animals by selective breeding, or what might be called artificial selection. So, Darwin's claim that drastic changes, even from one species to another, could be brought about over long ages of time by natural selection seemed reasonable. Unfortunately, the scientific community was also won over to Darwinism. Up to that time, scientists had pointed to the innumerable evidences of design in nature. These evidences pointed to a great

Creator.[2] The Bible says in Romans 1:19, 20 "Because that which may be known of God is manifest in them: for God hath shewed it unto them. For the invisible things of him from the creation of the world are clearly seen, being understood by the things that are made, even his eternal power and Godhead: so that they are without excuse." concerning Romans 1:20, Dr. Whitcomb wrote "the more we learn of the astronomic universe, the more we realize that evolutionism, even theistic evolutionism, offers no rational answers. Thus, in our generation, even more than in ancient times, God's eternal power and divine nature have been clearly seen."[3]

Darwin's famous book, *Origin of Species*, was probably aimed primarily at the scientific community. Referring to evidence of design in nature, Morris and Morris say it "would need to be explained by some other means, some naturalistic means, before evolution could really become acceptable to most people. And such a new explanation would need to be a 'scientific' explanation, or sufficiently so to convince the scientific community that it would really explain and confirm evolution."[4]

Darwin claimed that this naturalistic means was natural selection. This is the way that scientists were largely "converted" from being God-honoring advocates of design in nature to being the dedicated disciples of evolution. Once they were won over to Darwinism scientists then, for the most part, influenced the public to accept evolution. Charles Darwin was not really a great scientific pioneer at all. He was most likely an opportunist, with the right connections, who advocated natural selection at just the right time. Now that we've examined the historical context of Darwinism, let's look at its historical impact.

Charles Darwin took the phrase "struggle for existence" (which appears in *Origin of Species*) from the economist, Malthus. Thomas Malthus used this phrase in his *Essay on the Principle of Population* which concerned the tendency for populations to outgrow their food supply. The concept of "survival of the fittest," as applied to the human population is called social Darwinism. Businessmen such as John D. Rockefeller and Andrew Carnegie used Darwinism to justify their pitiless striving for wealth and power. According to Darwinism, it is natural for the inferior to be exploited by those who are superior. Social Darwinism basically equates wealth and power with genetic superiority. So, those who had influence and great fortunes could exploit other human beings, and the environment, and just sort of shrug it off as being the natural order of things. Thus, Social Darwinism seriously undermines the American values of equality and unalienable rights. It also stands in direct opposition to Biblical principles.

There is a close connection between Darwinism and racism. Darwinism gave racism a scientific justification, supposedly. The full title of Darwin's famous book was *The Origin of Species by Natural Selection or the Preservation of Favoured Races in the Struggle for Life.* Charles Darwin meant for his principles to apply to animal races (i.e. subspecies) as well as to the races of mankind. Darwin actually predicted the future eradication of, what he called, the "savage races" by the spread of the civilized Caucasian. He believed that the different races of *Homo sapiens* represented different degrees of evolution. Darwin considered the Caucasian race to be the highest stage of evolution that mankind had achieved. To Charles Darwin, the Black people represented the lowest evolutionary distance from apes. This same belief was also held by Thomas Huxley, who was probably the most important early advocate of Darwinism. These racist attitudes were almost universal among the evolutionists of the nineteenth century.

Dr. Morris wrote that, "evolutionary racism was a natural inference from the slow-and-gradual chance evolutionary process envisioned by Darwin and his followers. On that basis, race is simply a 'subspecies', which left to struggle for its existence in competition with other subspecies, may eventually triumph and become a distinct species."[5] Evolutionist anthropologists of the latter nineteenth century were largely responsible for the idea of Aryan superiority, which was used as a justification for racism in both Nazi Germany and South Africa. The term Aryan really means the parent language of a group of languages known as the Indo-European family. However, Aryan was mistakenly used to designate a racial group. This use was based on the assumption that those ethnic groups who spoke related languages must have a common racial source. This term came to be associated primarily with those of Anglo-Saxon heritage.

The concept of race is not really found in the Bible. Race is an evolutionary term which refers to a subspecies. Unfortunately, some have mistakenly interpreted the Bible to have racist connotations. The Bible says in Acts 17:26 that God " hath made of one blood all nations of men for to dwell on all the face of the earth." All people, no matter where they live, are "of one blood" in the sense that everyone is a direct descendent of Noah. So in God's eyes the only race is the human race.[6] May we all do our best to live up to the example of our great progenitor Noah, who "walked with God" (Genesis 6:9).

Evolutionary racism reached a terrifying extreme in Nazi Germany. In fact, Hitler made social evolution a basis for Germany's national policy. The evolutionist philosopher Frederick

Adolph Hitler

Nietzche had already popularized the concept of the master race and the individual "superman". Nietzche was a contemporary of Darwin and his ideas served as a foundation for Hitler's success.[7] Nietzche's philosophy had a far reaching influence. In Germany it contributed to racism and also militarism. Because Nietzche thought that Darwinism provided proof against the existence of God he announced that "God is dead." Nietzche died insane.[8]

Referring to Adolph Hitler, Paul Humber wrote that "he put survival-of-the-fittest into action and millions of 'unfit' people died as a result."[9] The book *Mein Kampf* had many references to evolution. For example, in Hitler's chapter titled "Nature and Race," he refers to the racial purity that he felt was necessary for the survival and development of the German people. It was Hitler's belief that it was a natural thing for one racial or ethnic group to try to establish its superiority over all others. Hitler seems to have been driven by his convictions about Teutonic racial superiority. Such a belief was, to some extent, based on Darwinism.

The national policy of Nazi Germany was based on evolution. The means by which Hitler sought to achieve racial purity was a process of systematic slaughter. Diabolical as the actions of Nazi Germany certainly were, these actions were consistent with what has been called evolutionary morality.[10] Morris and Morris wrote that Hitler. "was the ultimate evolu-tionist, if ever there was such a person. But he also was backed by the evolutionary scientists of Germany, who had become convinced followers of Ernst Haeckel and Charles Darwin." Ernst Haeckel was an evolutionary biologist who tried to directly transfer the "struggle for life" concept from evolutionary biology to the realm of politics and sociology. Haeckel was a major advocate of racism in that he believed that Germans were a geneti-cally superior group.[11] In his form of social Darwinism, evolutionist prin-ciples were to be proven politically, perhaps militarily. Nazism has been called "the ultimate fruit of the evolutionary tree."

The evolutionist scientific community played an important role in the atrocities of Nazi Germany. For example, Dr. Joseph Mengele, who was head of the Auschwitz death camp had impressive scientific credentials. He held both the MD and PhD degrees from highly respected German univer-sities. Because of his Darwinist background, with it's emphasis on survival of the fittest, Mengele believed that some individuals should not be allowed to reproduce. This policy was based on the fear that if a person had certain disor-ders or defects, he or she might pass them on to following generations.

Mengele and other Nazis believed that the best way to prevent certain people from passing on genetic defects, and thus achieve "racial purifica-tion" was to simply exterminate them. This policy was expanded to include all those without a "pure Teutonic genealogy", which included Jews, Gyp-sies, and blacks. The Nazis viewed these ethnic groups as less than human. It is estimated that at least eight million people died or were executed in

Hitler's concentration camps. Dr. Ben David Lew, who survived Buchenwald, wrote "people sent to Buchenwald were sent to die. The camp was filled with political vigilantes, the mentally retarded, Jehovah's Witnesses, homosexuals, physically handicapped persons, the elderly, but most of all, Jews. Thousands of Jewish people were sent there to fulfill Hitler's scheme of 'ridding Europe of the most terrible vermin'. The crematoriums were operated on a twenty-four hour basis."[12]

This is not to say that Charles Darwin advocated genocide. Dr. Morris wrote that "ideas do have unforeseen consequences. Darwin's idea that evolution means 'the preservation of favored races in the struggle for life' eventually led to Nazism and the Jewish holocaust even though Darwin himself would have been appalled at the thought."[13]

As with the Nazis, there was a strong connection between evolution and Communism. However, communists did not limit themselves strictly to Darwinism. Depending on the circumstances, communists also endorsed other versions of evolution such as Lamarckianism and punctuationism. Marxism needed the doctrine of evolution because it gave a kind of scientific respectability to atheism. Atheism removes the sacredness of human life by denying the existence of God, who sanctifies life. If people are merely animals, then they are expendable when the interests of "the people" would be served. The people's interests really means the political agenda of the ruling intellectual elite. Karl Marx was so delighted with Darwinism that it was his desire to dedicate part of *Das Kapital* to Charles Darwin. However, Darwin refused.[14]

Henry Morris also wrote, "Marx and Engels based their communistic philosophy squarely on the foundation of evolutionism. Their brand of evolution, however, was not pure Darwinism. With communism's emphasis on environmental influences, there has also been a long continued mixture of Lamarkianism (inheritance of acquired characteristics) and saltationism (sudden evolution)."[15] Lamarckianism gave the communist leadership God-like power in that they were supposedly able to alter history both politically and biologically through their programs and policies. In fact, the teaching of Mendelian genetics was actually outlawed by the communist leadership in 1948, in Russia.[16] Stalin was greatly influenced by Darwin. It has been documented that Joseph Stalin became an atheist as a schoolboy after reading *The Origin of Species*.[17] Marx and Engels were evolutionists, and all communists have been ever since. Evolution is actually the number one tenet of Marxism.[18]

Most people will probably be surprised to learn that there is a connection between evolution and abortion. Abortionists deny that a fetus is human by saying that the fetus goes through stages of progressive development. They claim that it is only in a late stage, near the end of the pregnancy, that the fetus has distinctly human qualities. As a matter of fact, this notion came from the German biologist, Ernst Haeckel. This concept was called the recapitulation theory. Haeckel published a series of drawings which ostensibly demonstrated that human and other mammalian embryos were identical. It was later found that these drawings were fraudulent.[19]

According to this recapitulation concept, the embryo goes through stages which reflect the supposed evolutionary development of that species. The late Carl Sagan was an advocate of recapitulation. Dr. Sagan claimed that the human embryo (then later the fetus) resembled a worm, a fish, a reptile, a pig, and a sub-human primate at successive stages of development.[20]

There are superficial similarities between the embryos of different organisms. Of course they all begin as just one cell. At one point in the development of the human embryo it has an attached sac which Haeckel said resembled a bird's yolk sac. However, this sac holds blood cells which are required for the embryo's growth. This little organ is very necessary and is not a vestige of some supposed earlier evolutionary stage. At another point in the embryo's development, linear folds develop which somewhat resemble gill slits in the embryo of a fish. Actual gills are used for breathing but these folds on the human embryo become necessary parts of the ear, the thymus gland, and the parathyrold gland. The structure on the human embryo that resembles a tail becomes the coccyx which is a bone that is essential for distinctly human movement and posture.[21]

Even though the "quasi-scientific notion" called recapitulation has been discredited, it is offered as a justification for abortion. Very little consideration is given to the rights of the unborn child because the fetus is not regarded as fully human. According to pro-choice doctrine, it is acceptable to terminate the pregnancy while the fetus is still in an animal stage of development. However, modern fetoscopy has proven that the unborn child is distinctly human in all stages of fetal development. Even Stephen Jay Gould, the prominent evolutionist, admits that recapitulation is now defunct.[22]

It has been scientifically verified that the unborn baby has the specifically human 46 chromosomes in his or her cells. The unborn baby is definitely alive at the instant of fertilization and is programmed for growth and development. His or her heartbeat is detectable at three weeks. In fact, the baby often has a different blood type than the mother. At the age of six weeks the baby has brain waves that are measurable on an electroencephalogram. The baby begins swallowing amniotic fluid at about eight weeks after fertilization. At nine weeks, ultrasound scanning allows parents to watch the movement of their baby inside the uterus. At eleven weeks the baby's hands are developed enough to grasp objects. Also, at eleven weeks, the baby's feet have been formed. He or she also has functioning organ systems, a skeletal structure, nerves, and circulation. The spinal cord and thalamus are developed at twelve weeks (or by the end of the first trimester). At fourteen weeks the heart is pumping vigorously. At 18 weeks, the baby is completely formed, and at 22 weeks is routinely viable outside of the womb.

About 1.5 million of these unmistakably human, unborn babies are killed by government sanctioned abortions in the United States each year. [23] Dr. Roy McLaughlin wrote that "four thousand three hundred little children will be murdered today while we are alive and breathing because we have freedom of choice! They will be killed without any type of trial and without any counsel to stand by their sides. They will be executed in a cruel and inhuman way."[24] Psalm 139, Psalm 5 1, Jeremiah I (among other passages of the Bible) make it clear that we are human beings from the moment of conception. Therefore, abortion can only be considered murder. As a matter of fact, Joseph Mengele, the "Angel of Death", performed abortions for a number of years in Buenos Aires. Paul Humber wrote that, "it should not be surprising that one who extinguished life at Auschwitz would practice a similar grisly crusade on life in the womb."[25]

Psychology is another area that has been strongly influenced by evolution. Morris and Morris wrote that "psychology, the study of the mind, is almost entirely based on the assumption that man is only an animal, derived by evolutionary descent from an ape-like ancestor. This was the view of Sigmund Freud, who has exerted probably the greatest single influence on the structure of modern psychology."[26] According to Morris and Morris, "Sigmund Freud is often listed together with Charles Darwin and Karl Marx as the three men whose teachings have had the greatest impact on the modern world. Furthermore, both Marx and Freud acknowledged their indebtedness to Darwin."[27] Freud had an obsessive hostility toward the Bible and Christianity, especially with regard to their moral restraints. Freud was an advocate of recapitulation and believed that a mental disorder was really a type of behavior that had been appropriate in a supposed previous stage of evolutionary development. Freud's psychological system was definitely atheistic.[28]

He saw sexual instinct as the impetus of behavior. Since the time of Freud, psychologists have considered man to be a highly evolved primate. The field of psychology has largely evaluated behavioral problems in an animalistic context. Many psychologists would even say that Christian beliefs are the result of a mental disorder. Evolutionists consider religion to be a vestige of societal pressures that existed within the groups of animals from which humans supposedly developed. Morris and Morris also wrote that "what was true of Freud became true of multitudes of his followers in succeeding generations, at least in their total rejection of the Christian faith and Biblical moral standards. They assume, erroneously, that Freud had disproved the validity of Christianity especially in view of the 'science' of evolution."

Evolutionary psychology is indirectly related to the sexual revolution. Psychologists and psychiatrists generally promote the idea that psychological problems are largely a result of sexual inhibitions. They reason that there is little sexual restraint observed among animals, so that should be considered normal for people too. After all, people are just higher animals (according to evolution). Traditional values such as chastity and fidelity are considered to be out of date, and perhaps rather ridiculous. Freud has been called the "grandfather of the modern sexual revolution."[29]

The field of education has also been influenced by evolution. Dr. Walter Wilson said that "our present state of child delinquency is largely due to the teaching in our schools of the evolutionary hypothesis and other anti-Christian doctrines. The result that naturally follows the teaching that human beings are a higher form of animal life is that our young people go out of the classroom to live like animals. They feel no responsibility at all to the God of Heaven."[30] It is probably fair to say that education is currently dominated by evolution in its content, philosophy, and methodology.

Modern educators have indoctrinated multitudes of students in evolutionary humanism. There has been a recent upsurge of opposition to the teaching of evolution in public schools. Christian leaders and concerned parents have begun to realize that a deadly anti-Biblical philosophy is being presented under the guise of science. This indoctrination takes place directly in courses such as geology and biology. It can also take place indirectly in the social sciences and even in the humanities. Evolutionist educators resent the fact that the validity of evolution is actually being questioned. Furthermore, they are vehemently opposed to the teaching of special creation in public, taxpayer supported institutions of learning.[31]

The event that may have triggered the takeover of American education by proponents of evolutionary humanism occurred when Charles W. Eliot became president of Harvard in 1869. Eliot was a Unitarian as well as an entrenched evolutionist. He appointed another Unitarian named John Fiske to be a professor of history and science. Fiske was the first person to teach evolution at Harvard, although Asa Gray, a botanist, taught evolution there at about the same time. Because of Harvard's prestigious leadership position, most other American colleges and universities soon began teaching evolution. Eventually, evolutionist doctrine even spread to public schools.

The most influential evolutionist, as far as American education is concerned, was a man named John Dewey. Dr. John N. Moore wrote that, "the 'prime mover' of modern education, John Dewey, showed a broad acceptance of Darwinism in his extensive writings. He viewed the human being as an 'evolved' creature that was slowly improving physically and mentally.[32] John Dewey was born in 1859, the same year that *The Origin of Species* was first published. Dewey was a student at Johns Hopkins University where he came under the influence of the evolutionist, James Hall. As a result of this association Dewey became, himself, an entrenched evolutionist. Dewey served as the head of teachers colleges at both the University of Chicago and Columbia University. Many educators were trained at these institutions that later assumed important leadership positions in American education. Dewey served as the first president of the American Humanist Association, which he helped to found in 1933. Humanism has an evolutionary/naturalistic explanation of origins as its philosophical basis. Humanism is an atheistic, man-centered philosophy.[33]

Humanists have done much to undermine the concept of the Christian home. An evolutionary origin of the universe has been called the first "tenet" of humanism. Humanists say that it is wrong to repress any form of sexual activity between consenting parties. They condemn fundamentalists and other religious people for trying to establish restraints on sexual conduct. Strong moral values are essential to the stability of the family. Since humanism is based on evolution it is not just a coincidence that the stability of the modern family has deteriorated as evolutionary theory has become universally accepted. Morris and Morris wrote that "evolutionism, with its premise that men and women are merely evolved animals, provides the perfect pseudo-scientific rationale for those who would do away with such scripture-based restraints." When man is viewed as an evolved animal, the concept of a Christian home becomes meaningless.[34]

It should be no great shock to anybody that an educational system based on humanism would eventually have no place for the Bible and prayer. It seems doubtful that much progress will be made in restoring the Bible and prayer to public education without addressing education's evolutionary/ humanistic philosophical basis. Doesn't it seem reasonable that a necessary forerunner of putting the Bible back in the public classroom is the teaching of creation science? Ken Ham wrote, "there are whole generations of students coming through an educational system who know nothing of the Bible. They have never heard about creation, Noah's Flood, or the message of the cross."[35]

Unfortunately, there has been considerable opposition to the balanced presentation of creation science along with evolution. There have been two highly controversial legal battles over laws that required the balanced treatment of creation and evolution. Both of these legal battles were a consequence of ACLU lawsuits which challenged their constitutionality.[36] The law that was passed by the state legislature in Arkansas was declared to be unconstitutional by a federal judge in Little Rock in 1981. The decision concerning the Arkansas law was never appealed. A similar law that was passed by the Louisiana legislature was also declared to be unconstitutional in a judgment by a federal judge in New Orleans. This Louisiana law was appealed to the Fifth Circuit Court of Appeals where a three judge panel supported the lower court's ruling. This decision was further appealed to the entire Fifth Circuit Court. These fifteen judges supported the previous decision eight to seven. This case was presented to the US Supreme Court in 1987. The nine justices upheld the previous decisions by a margin of seven to two, with William H. Rehnquist and Antonin Scalia dissenting.[37]

These decisions indicate that most modern judges have a decided bias in favor of evolution. However, they usually hide their biases behind the bogus principle of separation of church and state. Morris and Morris write that, "separation of church and state, incidentally, has never been a constitutional requirement, only a modern judicial interpretation. The First Amendment prohibited an established religion, which, in the context of the time, meant only the state endorsement and the support of a particular sect or denomination. It was certainly never intended to ban God from the schools and to establish the religion of secular humanism in our schools, as has been done."[38]

Many people in the legal profession even view the US Constitution as an evolving document. They reason that since there are no absolutes in an evolving universe, the US Constitution is an evolving document that must change with the times. The US Constitution was actually based on eternal, Biblical principles. Political science professors at the University of Houston examined the writings of the Founding Fathers. They carefully examined over 3,000 documents that were the most important in terms of their political impact. They found that these documents were primarily based upon the Bible. In fact 34% of the quotes in these documents were directly from the Bible. James Madison, who helped frame the US Constitution, said "we have staked the future of all our political institutions upon the capacity of each and all of us to govern ourselves according to the Ten Commandments of God." An 1892 Supreme Court decision said "Our laws and our institutions must necessarily be based upon, and embody, the teachings of the Redeemer of Mankind. It is impossible that it should be otherwise. To this extent our civilization and our institutions are emphatically Christian."[39]

Some creationists feel that the "creation laws" of Arkansas and Louisiana were rather ill-conceived anyway. This is because they were both compulsory laws which would basically have forced teachers to teach creation science. Enforcement of these laws would have been difficult. In most cases, teachers would not have had an adequate knowledge of the basic points of creation science. So the potential that these laws would have resulted in the dissemination of much misinformation about creation science was enormous.[40] It may be that creation laws should be constructed to permit, rather than compel, the teaching of the creation model of origins. A statute passed by the Kentucky legislature is just such a law. It ensures, or affirms the right of a public school teacher to teach the "Bible theory of creation." A portion of KRS 158.177 says that, "In any public instruction concerning the theories of the creation of man and earth, and which involves the theory thereon commonly known as evolution, any teacher so desiring may include as a portion of such instruction the theory of creation as presented in the Bible, and may accordingly read such passages in the Bible as are deemed necessary."

The most famous trial concerning a creation law was the Scopes Trial in 1925. This trial took place at the courthouse in Dayton, Tennessee. Most of the more than 200 news reporters that covered this trial were very biased against a literal interpretation of Genesis. This trial took some odd turns and was said to have had a circus-like atmosphere. Perhaps the strangest thing about this trial was that the defense asked the jury to find John T. Scopes guilty. John Scopes probably never taught evolution in a public school. William Jennings Bryan, the great fundamentalist statesman, was actually the central figure of the trial.[41]

The law in question was an "anti-evolution law" which had recently been passed by the Tennessee state legislature. The American Civil Liberties Union wanted to test this law in court and offered to pay the expenses of any teacher who was willing to cooperate. An engineer with the Cumberland Coal and Iron Company, named George Rappleyea, saw an ad about the ACLU's offer in The Chattanooga Times. Then Rappleyea consulted with local officials there in Dayton. They found a young teacher named John Scopes to play the role of the "law-breaking villain." These Dayton officials really meant the trial to be a sort of a publicity stunt to generate interest in the industrial development of their town.[42]

However, the leadership of the ACLU, far away in New York City, was pursuing an entirely different agenda. They viewed this as a chance to destroy fundamentalism by ridiculing it in the media. At that time, fundamentalist Christians were generating a tremendous anti-evolution sentiment across the country. This was especially true with regard to the teaching of evolution in schools. The famous attorney, Clarence Darrow, was chosen to defend John Scopes in the trial. Darrow was an agnostic who was said to have "found Bryan's opinions dangerous."[43] Darrow had made a name for himself as a labor lawyer and also as the defense attorney in many notable murder trials.

William Jennings Bryan was chosen to lead the prosecution in the Scopes trial. Bryan was a former Secretary of State and had run for President. He had also represented Nebraska in the US Congress. Bryan lost by a narrow margin to McKinley in the 1896 presi-

William Jennings Bryan

dential election. He also received the Democratic nomination to run for president in 1900 and 1908.

William Jennings Bryan was a friend of the evangelist Billy Sunday and strongly supported the cause of Prohibition. Bryan believed in and defended the idea of an inerrant, literal Bible. He was very disturbed about the effect the teaching of evolution was having on students, in terms of their beliefs and morality. Bryan preached against evolution across the nation.[44] These lectures were syndicated and appeared in many of America's newspapers. One such lecture, titled, "Is the Bible True?" had, itself, a wide distribution in printed form. This lecture was said to have led to the passage of Tennessee's Butler Act. This law prohibited the teaching of evolution in all of Tennessee's public schools. [45]

The actual trial was held in July of 1925. Perhaps the primary goal of the ACLU leadership at that point was to undermine the powerful influence of William Jennings Bryan. During that trial Bryan actually took the witness stand himself and was questioned by Darrow. Clarence Darrow was trying to make Bryan look ridiculous on the stand. For example, Bryan made a serious error when he expressed confidence in the inerrancy of the Bible, but also advocated the day/age theory. Of course, Darrow pounced on this inconsistency. He argued that it was ludicrous to say that Scripture was infallible and yet was open to such a flexible interpretation. As far as this particular point is concerned, Darrow was absolutely right.[46]

On the eighth day of the trial, the extensive exchange between Darrow and Bryan was removed from the record. This was largely due to the fact that many of Darrow's questions were rather silly and beside the point (of teaching evolution). When Darrow recognized that he couldn't clear the defendant, he actually asked the judge to give instructions to the jury to render a guilty verdict. The jury did so and Scopes was fined $100, Upon appeal, the Supreme Court of Tennessee confirmed the constitutionality of the Butler Act. Sadly, that outstanding fundamentalist, William Jennings Bryan, died five days after the Scopes trial. Bryan was 65 years old when he died.

The conviction of Scopes was really a pyrrhic victory. Dr. John Morris wrote that "even though John Scopes was 'convicted' of teaching evolution, in many ways the anti-creationists won a major victory, since to the victor belongs the spoils. During the trial, Christians were depicted as ignorant, foolish anti-intellectuals. And this image was portrayed to the world by the media. Evolutionists are still enjoying the incredible gains made through the Scopes Trial."[47]

There is a much more recent historical development that is related to evolution. In the previous chapter, we saw that humanists attempt to exalt human reason above Biblical authority. Dr. Henry Morris wrote that "humanism over the ages has taken many different forms, and some of the ancient faiths are making a remarkable resurgence today in a modern pseudo-scientific garb. Many people, including a considerable number of scientists, are perceptive enough to see that random processes in a chance universe could never generate the multitude of highly complex systems that now exist in the universe. However, being unwilling to attribute these systems to the personal Creator God of the Bible, they assume that the cosmos itself has creative powers, or even that it is alive and directing its own evolution." This ancient belief is being revived by the "so-called" New Age movement.[48]

The New Age Movement is a far-reaching, worldwide conspiracy. It is actually an attempt to convert much of the world, including the United States, to a different religion. This is a religion that denies the deity of Christ. This new religion is closely related to the concept of a New World Order. Very influential people within the New Age Movement are trying to establish a unified, global leadership that is both religious and political. That is what the phrase New World Order really refers to. Dr. Morris wrote "the immense variety of organizations and sub-movements that are part of the New Age complex would almost defy analysis except for the fact that all are based on evolutionism, as viewed in a pantheistic context, with this evolutionary process now operating on the human level and aimed at an eventual world culture under a world government."[49]

Many Christians believe that this New Age Movement will eventually usher in a "one-world" government under the leadership of the Antichrist. According to Dr. Sexton, "there is a master conspiracy, someone obeying and carrying out the plan, sometimes unknowingly acting for the Devil. Read for yourselves the Devil's goal in the last book of the Bible. It is a one-world political order and a one-world religion led by the Antichrist and the False Prophet. That's the goal of Satan and we are seeing it unfold before our eyes."[50]

It is probably fair to say that Marilyn Ferguson is one of the most important advocates of the New Age religion. She wrote a book called *The Aquarian Conspiracy*. This book has been referred to as the "Bible" of the New Age Movement. Marilyn Ferguson actually has much to say about evolution in this influential book. New Age philosophy embraces the punctuated equilibrium model of evolution. She wrote in *The Aquarian Conspiracy* that "Darwin's theory of evolution by chance mutation and survival of the fittest has proven hopelessly inadequate to account for a great many observations in biology," She went on to say that "to this day fossil evidence has not turned up the necessary missing links. Gould called the extreme rarity in the fossil record of transitional forms of life 'the trade secret of paleontology.' Younger scientists, confronted by the continuing absence of such missing links, are increasingly skeptical of the old theory."

As previously mentioned, the punctuated equilibrium model basically says that there are highly accelerated periods of evolution. Marilyn Ferguson also wrote that "Gould and Eldredge independently proposed a resolution of this problem, a theory that is consistent with the geological record. Soviet paleontologists have proposed a similar theory. Punctuationalism or punctuated equilibrium suggests that the equilibrium of life is 'punctuated' from time to time by severe stress. If a small segment of the ancestral population is isolated at the periphery of its accustomed range, it may give way to a new species. Also, the population is stressed intensely because it is living at the edge of its tolerance."[51]

Marilyn Ferguson believes that the significance of the punctuated equilibrium model is that it means that there is a possibility that an accelerated period of human evolution may occur in the near future. She believes that psychic phenomena (e.g., telepathy, precognition, and psychokinesis) may be evidence that human evolution is about to take place. Marilyn Ferguson wrote that "millions are experimenting with the psychotechnologies. Are they creating a more coherent, resonant society, feeding order into the great social hologram like seed crystals? Perhaps this is the mysterious process of collective evolution."[52]

New Agers believe that mankind will someday evolve into a state of "higher consciousness." Texe Marrs wrote that "a recent and profound addition to the theory of evolution is the concept *of punctuated equilibrium,* first articulated in the Soviet Union in the early 1970's. According to this theory, species can develop very quickly. It does not take millions of years of natural selection, as Darwin proposed, for a new species to evolve."

According to Marrs, "New Age teachers and gurus are fascinated by the idea that modern man may be on the precipice of an incredible evolutionary leap — to a level of superhuman higher consciousness. He will then be a god."[53]

In conclusion, evolutionary theory has had a profound historical impact. Evolution can be likened to a poisonous plant which has sent its tendrils virtually everywhere. Darwinism in particular has been a philosophical basis for Communism, Nazism, racism, humanism, abortion, and even the sexual revolution. It is clear that Charles Robert Darwin cast a long and very sinister shadow across the course of modern history.

Literature Citations: Chapter 1

1. Morris, H.M., *The Long War Against God*, Baker, p. 23-25
2. Ham, Ken, *The Lie: Evolution*, Master, p. 5
3. Ham, Ken, *The Lie: Evolution*, Master, p. 3
4. Chittick, D.E., *The Controversy*, Multnomah, p. 20
5. Morris, H.M. and G.E. Parker, *What Is Creation Science?*, Master, p. 8-9
6. Gish, D.T., *Teaching Creation Science In Public Schools*, Institute for Creation Research (ICR), p. 31-32
7. Bryan, W.J., *The Bible Or Evolution*, Sword of the Lord Foundation, p. 17
8. Gish, D.T., *Evolution: The Challenge of the Fossil Record*, Master, p. 12-15
9. Ham, Ken, *The Lie: Evolution*, Master, p. 44
10. Morris, H.M., *Biblical Creationism*, Baker, p. 228-32
11. Morris, H.M., and J.C. Whitcomb, *The Genesis Flood*, Presbyterian and Reformed, p. 459-465
12. Chittick, D.E., *The Controversy*, Multnomah, p. 178
13. Huse, S.M., *The Collapse of Evolution*, Baker, p. 4
14. Whitcomb, J.C., *The Early Earth*, Baker, p. 119
15. Morris, H.M. and J.C. Whitcomb, *The Genesis Flood*, Presbyterian and Reformed, p. 457
16. Morris, H. M., *Twilight of Evolution*, Baker, p. 23
17. Ham, Ken, *The Lie: Evolution*, Master, p. 25
18. Talmage, T.D., *Sword of the Lord*, 4 Aug. 2000, p. 20

Literature Citations: Chapter 2

1. DeYoung, D.B., *Astronomy and the Bible*, Baker, p. 120
2. Rice, Bill, "Science and the Bible," Sword of the Lord, 17 May 1996, p. 14
3. Morris, H.M., *Science and the Bible*, Moody, p. 14
4. Rice, Bill, "Science and the Bible," Sword of the Lord, 17 May 1996, p.14
5. Morris, H.M., *Science and the Bible*, Moody, Chicago, 1986, p.14
6. Morris, H.M., *Science and the Bible*, p. 18-19
7. Davies, Keith, *Evidences for a Young Sun*, Impact no. 276, ICR, June, 1996, p. 1-3
8. *Compton's Pictured Encyclopedia*, F.E. Compton and Company, Chicago, Vol. 13, page 514
9. Morris, H.M., *Science and the Bible*, Moody, p.19
10. Morris, H.M., *Science and the Bible*, Moody. p. 21-23
11. Gilson, Etienne, *History of Christian Philosophy in the Middle Ages*, Random House, p. 368-371
12. Henry Matthew, *Commentary on the Whole Bible*, Zondervan, p. 75
13. McLaughlin, J.F., *Abingdon Bible Commentary*, Abingdon, p. 56
14. Hindson, E.E., and W.M. Kroll, *KJV Parallel Bible Commentary*, Thomas Nelson, p 24
15. Spurgeon, C.H., *Spurgeon's Devotional Bible*, Baker, p. 74
16. Morris, H.M., *Biblical Basis for Modern Science*, Baker, p. 34-36
17. Muller, Richard, *Great Thinkers of the Western World*, Harper-Collins, p. 109

18. Morris, H.M., *Biblical Basis for Modern Science*, p. 36-37
19. Morris Henry, *Biblical Basis for Modern Science*, p. 190
20. Brown, Laurie M. and R.T. Weidner, "Physics," *Encyclopedia Britannica*, Vol. 25, p. 832
21. Huse, S.M., *Collapse of Evolution*, Baker, p. 61
22. Gish, Duane, *Creation Scientists Answer Their Critics*, ICR, p. 153
23. Morris, H.M. and Gary Parker, *What Is Creation Science?*, Master, p. 256-59
24. Gish, Duane, *Creation Scientists Answer Their Critics,* ICR, p. 151
25. Morris, H.M., *Biblical Basis of Modern Science*, Baker, p. 187
26. Rice, J.R., *In The Beginning*, Sword of the Lord, p. 22

Literature Citations: Chapter 3

1. Rice, J.R., *Evolution or the Bible*, Sword of the Lord, p. 4
2. Gish, Duane, *Evolution: Challenge of the Fossil Record*, Master, p. 11
3. Johnson, Phillip, *Darwin On Trial*, Regnery Gateway, p. 151
4. Huse, S.M., *The Collapse of Evolution*, Baker, p. 146
5. Kevles, Bettyann, *Encyclopaedia Britannica*, Encyclopaedia Britannica Inc., Vol. 16, p. 977-978
6. Huse, S.M., *The Collapse of Evolution*, Baker, p. 9
7. Johnson, P.E., *Darwin on Trial*, Regnery Gateway, p. 16-17
8. Darwin, Charles, *The Origin of Species*, Mentor, p. 91-92
9. Chittick, Donald, *The Controversy*, Multnomah, p. 67
10. Ruse, Michael, *Darwinian Revolution*, University of Chicago, p. 178-179
11. Johnson, P.E., *Darwin on Trial*, Regnery Gateway, p. 18-19
12. Gish, Duane, *Evolution: Challenge of the Fossil Record*, Master, p. 37-38
13. Macomber, Richard W., *Britannica,* Encyclopaedia Britannica, Inc., Vol.7, p. 585-586.
14. Rice, J.R., *In The Beginning*, Sword of the Lord, 1975, p. 56
15. Nicolle, Jacques, *Britannica,* Encyclopaedia Britannica, Inc., Vol. 25, p. 455
16. Gish, Duane, *Teaching Creation Science In Public School*, ICR, p. 11-12
17. Whitcomb, John C. and Henry Morris, *The Genesis Flood* Presbyterian and Reformed, p. 459-464
18. Morris, H.M. and Gary Parker, *What is Creation Science?*, Master, p. 199
19. Huse, S.M., *The Collapse of Evolution*, Baker, p. 64
20. Whitcomb, John C. and Henry Morris, *The Genesis Flood*, Presbyterian and Reformed, p. 459-464
21. Whitcomb and Morris, *The Genesis Flood,* p. 95-96
22. Huse, S.M., *The Collapse of Evolution*, Baker, p. 153
23. Ruse, Michael, *The Darwinian Revolution*, University of Chicago, 1979, p 49-51.
24. Huse, S.M., *The Collapse of Evolution*, Baker, p. 143
25. Morris, H.M., *The Long War Against God*, Baker, p. 168
26. Whitcomb and Morris, *The Genesis Flood*, p. 90
27. Rice, J.R., *In The Beginning*, Sword of the Lord, p. 46
28. Rice, J.R., *Evolution or the Bible,* Sword of the Lord, p. 28-30
29. Morris, H.M., *Twilight of Evolution*, Baker, p. 75-76
30. Morris, H.M., *The Long War Against God*, Baker, p. 199
31. Morris, H.M., *The Long War Against God*, Baker, p. 185-186

32. Lovejoy, A.O., *The Great Chain of Being* , Harper and Row, p. 15
33. Morris, H.M., *Long War Against God*, p. 204
34. Lovejoy, A.O., *The Great Chain of Being*, Harper and Row, p. 89
35. Lovejoy, A.O., *The Great Chain of Being*, Harper and Row, p. 60-61
36. Lovejoy, A.O., *Great Chain of Being*, p. 80
37. Lubenow, Marvin, *Bones of Contention*, Baker, p. 95
38. Morris, H.M., *The Long War Against God*, Baker, p 187
39. Lubenow, Marvin, *Bones of Contention*, Baker, p. 93-94
40. Morris, H.M., *The Long War Against God*, Baker, p. 184
41. Huse, S.M., *The Collapse of Evolution*, Baker, p. 88
42. Hitching, Francis, *The Neck of the Giraffe*, Meridian, p. 121
43. Parker, G.E. and H.M. Morris, *What Is Creation Science?*, Master, p. 92
44. Morris, H.M., *The Long War Against God*, Baker, p. 157
45. Morris, H.M., *The Long War Against God*, p. 177
46. Morris, H.M., *The Long War Against God*, p. 157
47. Parker, Gary and H.M. Morris, *What Is Creation Science?*, Master, p. 82
48. Hitching, Francis, *The Neck of the Giraffe*, Meridian, p. 199
49. Morris, H.M., *The Long War Against God*, Baker, p. 174
50. Hitching, Francis, *The Neck of the Giraffe*, Meridian, p. 213
51. Morris, H.M., *The Long War Against God*, Baker, p. 167
52. Morris, H.M., *The Long War Against God*, Baker, p. 242

Literature Citations: Chapter 4

1. Morris, H.M., *Men of Science — Men of God*, Master, p.1
2. Chittick, Donald, *The Controversy*, Multnomah, p. 16-17
3. Morris, H.M., *The Long War Against God*, Baker, p. 304-305
4. Chittick, Donald, *The Controversy*, Multonomah, p.17
5. Morris, H.M., *The Biblical Basis for Modern Science*, Baker, p. 29
6. Huse, S.M., *The Collapse of Evolution*, Baker, p. 118
7. Morris, H.M., *Men of Science — Men of God*, Master, p. 9-11
8. Morris, H.M., *Men of Science — Men of God*, Master, p. 11-13
9. McGreal, Ian, editor, *Great Thinkers of the Western World*, Harper-Collins, p. 182
10. Morris, H.M., *Men of Science — Men of God*, Master, p. 13
11. McGreal, Ian, editor, *Great Thinkers of the Western World*, Harper-Collins, p.172-173
12. Johnson, Phillip, *Darwin on Trial*, Regnery Gateway, p. 146
13. Morris, H.M., *Men of Science — Men of God*, Master, p. 15
14. Krailsheimer, A.J, *Encyclopedia Americana*, Vol. 1, Grolier, p. 504
15. Morris, H.M., *Men of Science — Men of God*, Master, p. 16
16. Morris, H.M., *Men of Science — Men of God*, Master, p. 16-18
17. Morris, H.M., *Men of Science — Men of God*, Master, p. 18
18. Morris, H.M., *Men of Science — Men of God*, Master, p. 27
19. Whitcomb, John C. and H.M. Morris, *The Genesis Flood*, Presbyterian and Reformed, p. 91
20. Morris, H.M., *Men of Science — Men of God*, Master, p. 26
21. Morris H.M., *Men of Science — Men of God*, Master, p. 27
22. Morris, H.M., *Men of Science — Men of God*, Master, p. 21

23. Morris H.M., *Men of Science — Men of God,* Master, p. 30-31
24. McGreal, Ian, editor, *Great Thinkers of the Western World*, Harper-Collins, p. 237-240
25. Morris, H.M., *Men of Science — Men of God*, Master, p. 31
26. Morris, H.M., *Men of Science — Men. of God*, Master, p. 29
27. Sword Calendar, 1997, Sword of the Lord
28. Morris, H.M., *Men of Science — Men of God*, Master, p. 29
29. Morris, H.M., *Men of Science — Men of God*, Master, p. 39-40
30. Morris, H.M., *Men of Science — Men of God*, Master, p. 37
31. Morris, H.M., *Men of Science —Men of God*, Master, p. 44
32. Morris, H.M., *Men of Science —Men of God*, Master, p. 47
33. Morris, H.M., *Men of Science —Men of God*, Master, p. 49
34. Morris, H.M., *Biblical Basis for Modern Science*, Baker, p. 290
35. Morris, H.M., *Men of Science — Men of God*, Master, p. 52
36. Huse, S.M., *The Collapse of Evolution*, Baker, p. 118
37. Morris, H.M., *Men of Science — Men of God*, Master, p. 53
38. Morris, H.M., *Men of Science — Men of God*, Master, p. 59
39. Morris, H.M., *Men of Science — Men of God*, Master, p. 60-62
40. Burchfield, J.D., *Encyclopedia Americana*, Vol. 16, Grolier, p. 354-355
41. Morris, H.M., *Men of Science — Men of God*, Master, p. 67
42. Morris, H.M., *Men of Science — Men of God*, Master, p. 72
43. Tisdale, Samuel and W.L. Nelson, *Soil Fertility and Fertilizers*, Macmillan, 1975, p. 15
44. Morris, H.M., *Men of Science — Men of God,* Master, p. 75
45. Morris, H.M., *Men of Science — Men of God*, Master, p. 77
46. McDowell, Josh, *Evidence That Demands a Verdict,* Here's Life Publishers, p. 71
47. Morris, H.M., *Men of Science — Men of God*, Master, p.80
48. Morris, H.M., *Men of Science — Men of God*, Master, p. 81-83
49. Sword Calendar, 1996, Sword of the Lord
50. Rice, J.R., *Strange Short Stories By The Doctor*, Sword of the Lord, p. 51
51. Morris, H.M., *Men of Science — Men of God*, Master, p. 85-88
52. Whitcomb, John C. and H.M. Morris, *The Genesis Flood*, Presbyterian and Reformed, p. 90

Literature Citations: Chapter 5

1. Woodmorappe, John, *Noah's Ark: A Feasibility Study*, ICR, p. XI
2. Whitcomb, John C. and H.M. Morris, *The Genesis Flood*, Presbyterian and Reformed, p. 10
3. Sightler, Harold B., *Genesis: Vol. 1*, Tabernacle Baptist Church, p. 80
4. Morris, H.M., *The Genesis Record*, 1993, Baker, p. 182
5. Hindson, E.E. and W.M. Kroll, *The KJV Parallel Commentary*, Thomas Nelson, p. 31
6. Whitcomb, John, *The World That Perished*, 1993, Baker, p. 22
7. Morris, H.M., *The Genesis Record*, 1976, Baker, p.
8. Hindson, E.E. and W.M. Kroll, *The KJV Parallel Commentary*, p. 32
9. Whitcomb, John C., *The World That Perished*, Baker, p. 25
10. Whitcomb, J. C. and H.M. Morris, *The Genesis Flood*, Presbyterian and Reformed, p. 69

11. Morris, H.M., *Science and the Bible,* 1986, Moody, p. 87
12. Woodmorappe, John, *Noah's Ark: A Feasibility Study,* ICR, p. 5-7
13. Woodmorappe, John, *Noah's Ark: A Feasibility Study,* ICR, p. 13
14. Hindson, E.E. and W.M. Kroll, *The KJV Parallel Commentary,* Thomas Nelson, p. 32
15. Sightler, Harold, *Genesis: Vol. 1,* Tabernacle Baptist Church, p. 89
16. Woodmorappe, John, *Noah's Ark: A Feasibility Study,* ICR, p. 91
17. Davis, J.D., professor of agriculture, Murray State University, personal interview, 11 May 98.
18. Woodmorappe, John, *Noah's Ark: A Feasibility Study,* ICR, p. 95-98
19. Woodmorappe, John, *Noah's Ark: A Feasibility Study,* ICR, p. 17-19
20. Woodnorappe, John, *Noah's Ark: A Feasibility Study,* ICR, 1996, p. 97
21. Woodmorappe, John, *Noah's Ark: A Feasibility Study,* ICR, p. 20
22. Woodmorappe, John, *Noah's Ark: A Feasibility Study,* ICR, p. 62-63
23. Woodmorappe, John, *Noah's Ark: A Feasibility Study,* ICR, p. 84
24. Rice, J.R., *In The Beginning,* 1975, Sword of the Lord, p. 196
25. Reed, Sherwood C., R.W. Crites, and E.J. Middlebrooks, *Natural Systems for Waste Management and Treatment,* McGraw-Hill, p. 355
26. Reed, Sherwood, C., R.W. Crites, and E.J. Middlebrooks, *Natural Systems for Waste Management and Treatment,* McGraw-Hill, p. 75
27. Woodmorappe, John, *Noah's Ark: A Feasibility Study,* ICR, p. 27-34
28. Woodmorappe, John, *Noah's Ark: A Feasibility Study,* ICR, p. 2
29. Woodmorappe, John, *Noah's Ark: A Feasibility Study,* ICR, p. 44
30. Woodmorappe, John, *Noah's Ark: A Feasibility Study,* ICR, p. 72-76
31. Ben-Lew, David, Clearwater, FL, personal communication, 1997
32. Rice, J.R., *In the Beginning,* 1975, Sword of the Lord, p. 229
33. Hutson, Curtis, *Salvation Crystal Clear, Vol. 1,* Sword of the Lord, p. 239
34. Greene, Oliver B., *Commentary on Ephesians,* Gospel Hour, Inc., p. 43
35. Rice, J.R., *In the Beginning,* Sword of the Lord, p. 229
36. Hutson, Curtis, *How To Know You Are Going to Heaven,* Sword of the Lord
37. Spurgeon, C.H., *Spurgeon's Devotional Bible,* Baker, p. 753
38. Hyles, J.F., "Worthy Is the Lamb" (audio cassette), Hyles Publications
39. Rice, J.R., *The Son of God,* Sword of the Lord, p. 380
40. Ben-Lew, David, *Passover — Festival of Freedom* (pamphlet), used by permission
41. Hankins, J.H., *Sword of the Lord,* 22 Dec. 2000, p. 22
42. Lowry, Robert, "Nothing But the Blood," *Baptist Hymnal,* Convention Press
43. Ben-Lew, David, "Yom Kippur — Day of Atonement" (pamphlet), used by permission
44. Spurgeon, C.H., *Spurgeon's Devotional Bible,* 1990, Baker, p. 109
45. Ben-Lew, David, "Yom Kippur — Day of Atonement" (pamphlet), used by permission
46. Sightler, Harold, *Sword of the Lord,* 21 Oct. 1994, p. 17
47. McDowell, Josh, *Evidence That Demands A Verdict,* Here's Life Publishers, p. 186-88
48. Torrey, R.A., *Sword of the Lord,* 17 Dec. 1993, p. 23
49. Ben-Lew, David, "Must Messiah Be Virgin Born?" (pamphlet)
50. McDowell, Josh, *Evidence That Demands A Verdict,* Here's Life Publishers, p. 112
51. Roloff, Lester, *Sword of the Lord,* 20 March 1998, Sword of the Lord, p. 20-25

Literature Citations: Chapter 6

1. Whitcomb, J.C. and H.M. Morris, *The Genesis Flood*, Presbyterian and Reformed, p. 9
2. Morris, H.M., *The Genesis Record*, Baker, p. 96
3. Whitcomb, J.C. and H.M. Morris, *The Genesis Flood*, Presbyterian and Reformed, p. 122
4. Huse, S.M., *Collapse of Evolution*, Baker, p. 33
5. Morris, H.M., *The Genesis Record*, Presbyterian and Reformed, p. 194-95
6. Whitcomb, J.C. and, H.M. Morris, *The Genesis Flood*, Presbyterian and Reformed, p. 121
7. Dillow, Joseph, *The Waters Above*, 1982, Moody, p. 137
8. Morris, H.M., *The Genesis Record*, Baker, p. 197
9. Whitcomb, J.C. and H.M. Morris, *The Genesis Flood*, Presbyterian and Reformed, p. 287
10. Dillow, Joseph, *The Waters Above*, Moody, p. 78.
11. Whitcomb, J.C. and H.M. Morris, *The Genesis Flood*, Presbyterian and Reformed, p. 243-245
12. Whitcomb, J.C., *The World That Perished*, Baker, p. 81
13. Morris, H.M., *Science and the Bible*, Moody, p. 78
14. Whitcomb, J.C. and H.M. Morris, *The Genesis Flood*, Presbyterian and Reformed, p. 268
15. *Compton's Pictured Enciclolpedia*, F.E. Compton and Co., Vol. 6, p. 61-62
16. Whitcomb, J.C. and H. M. Morris, *The Genesis Flood*, Presbyterian and Reformed, p. 268
17. Ham, Ken, and Snelling A., Wieland, C., *The Answers Book*, Master, p. 126-29
18. Whitcomb, J.C. and H. M. Morris, *The Genesis Flood*, Presbyterian and Reformed, p. 267
19. Rice, J.R., *In The Beginning*, 1975, Sword of the Lord, p. 251
20. Whitcomb, J.C. and H. M. Morris, *The Genesis Flood*, Presbyterian and Reformed, p. 292-94
21. Dillow, Joseph, *The Waters Above*, 1982, Moody, p. 187
22. Ham, Ken, and Snelling A., Wieland, C., *The Answers Book*, Master, p. 77-79
23. Whitcomb, J.C., *The World That Perished*, Baker, p. 86
24. Vardiman, Larry, ICR Impact #277, July 1996
25. Wright, J.W., editor, *The Greenhouse Effect*, The Universal Almanac, Universal Press Syndicate, p. 369
26. Rice, J.R., *In The Beginning*, 1975, Sword of the Lord, p. 245-46
27. Scofield, C.I., *The Scofield Reference Bible*, 1945, Oxford, p. 19.
28. Whitcomb, J.C. and H. M. Morris, *The Genesis Flood*, Presbyterian and Reformed, p. 86.
29. Ham, Ken, and Snelling A., Wieland, C., *The Answers Book*, Master, p. 85.
30. Scofield, C.I., *The Scofield Reference Bible*, Oxford, p. 569
31. Morris, H.M., *The Remarkable Record of Job*, Baker, p. 29-30
32. Dillow, Joseph, *The Waters Above*, 1982, Moody, p. 186
33. Dillow, Joseph, *The Waters Above*, Moody, p. 190
34. Dillow, Joseph, *The Waters Above,* Moody, p. 182-83
35. Parker, Gary, *What is Creation Science?*, Master, p. 168

36. Morris, H.M., *Science and the Bible*, Moody, p. 79-80
37. Morris, H.M., *The Genesis Record*, Baker, p. 204
38. Dillow, Joseph, *The Waters Above*, Moody, p. 183
39. Huse, S.M., *Collapse of Evolution*, Baker, p. 46
40. Whitcomb, J.C., *The World That Perished*, Baker, p. 49
41. Morris, H.M., *The Genesis Record*, Baker, p. 201
42. Hindson, E.E.and W. M. Kroll, *The KJV Parallel Bible Commentary*, Thomas Nelson, p. 33
43. Morris, H.M., *The Genesis Record*, Baker, p. 199

Literature Citations: Chapter 7

1. Levin, H.L. *The Earth Through Time*, 5th edition, Saunders, p. 53
2. Levin, H.L. *The Earth Through Time*, Saunders, p. 73
3. Whitcomb, J.C. and H.M. Morris, *The Genesis Flood*, Presbyterian and Reformed, p. 124
4. Whitcomb, J.C. and H.M. Morris, *The Genesis Flood*, Presbyterian and Reformed, p. 144
5. Huse, S.M. *The Collapse of Evolution*, Baker, p. 153
6. Gish, Duane, *Evolution: The Challenge of the Fossil Record*, Master, p. 46
7. Gish, Duane, *Evolution: The Challenge of the Fossil Record*, 1985, Master, p. 47
8. Huse, S.M. *The Collapse of Evolution*, Baker, p. 14-15
9. Austin, S.A., ICR Impact #137, November 1984
10. Whitcomb, J.C. and H.M. Morris, *The Genesis Flood*, Presbyterian and Reformed, p. 208-209
11. Whitcomb, J.C., *The World That Perished*, Baker, p. 86, 87
12. Whitcomb, J.C. and H.M. Morris, *The Genesis Food*, Presbyterian and Reformed, p. 181
13. Whitcomb, J.C. and H.M. Morris, *The Genesis Flood*, Presbyterian and Reformed, p. 171
14. Levin, H.L. *The Earth Through Time*, 5th edition, Saunders, p. 78
15. Whitcomb, J.C. and H.M. Morris, *The Genesis Flood* , Presbyterian and Reformed, p. 171
16. Huse, S.M. *The Collapse of Evolution*, Baker, p. 33-35
17. Gish, D.T., *Dinosaurs By Design*, 1992, Master, p. 8
18. Meldau, F.J., *Why We Believe In Creation Not In Evolution*, Christian Victory Publishing, p. 321
19. Velikovsky, Immanuel, *Earth In Upheaval*, Doubleday, p. 221-222
20. Whitcomb, J. C. and H.M. Morris, *The Genesis Flood,* Presbyterian and Reformed, p. 275
21. Whitcomb, J.C. and H.M. Morris, *The Genesis Flood*, Presbyterian and Reformed, p. 123
22. Huse, S.M. *The Collapse of Evolution*, Baker, p. 46-47
23. Whitcomb, J. C. and H. M. Morris, *The Genesis Flood* , Presbyterian and Reformed, p. 273-276
24. Meldau, F.J., *Why We Believe In Creation Not In Evolution*, Christian Victory Publishing, p. 311
25. Gish, D.T., *Creation Scientists Answer Their Critics,* ICR, p. 115-18

26. Chittick, D.E., *The Controversy*, Multnomah, p. 63-64
27. Gish, D.T., *Evolution: The Fossils Still Say No!*, ICR, p. 56
28. Morris, H.M. and J.D. Morris, *Science and Creation*, Master, p. 282
29. Morris, H.M. and J.D. Morris, *Science and Creation*, Master, p. 60
30. Johnson, Phillip, *Darwin on Trial*, Regnery Gateway, p. 54-55
31. Gish, D.T., *Creation Scientists Answer Their Critics*, ICR, p. 127
32. Gish, D.T., *Evolution: The Fossils Still Say No!*, ICR, p. 76
33. Huse, S.M., *The Collapse of Evolution*, Baker, p. 44-45
34. Gish, D.T., *Evolution: The Fossils Still Say No!*, ICR, p. 83-87
35. Morris, H.M. and J.D. Morris, *Science and Creation*, Master, p. 64
36. Gish, D.T., *Evolution: The Fossils Still Say No!*, ICR, p. 96-98
37. Johnson, Phillip, *Darwin On Trial*, Regnery Gateway, p. 75
38. Gish, D.T., *Evolution: The Fossils Still Say No!*, ICR, p.111-114
39. Gish, D.T., *Dinosaurs By Design*, 1992, Master, p. 58
40. Gish, D. T., *Evolution: The Fossils Still Say No!*, ICR, p. 103
41. Huse, S.M. *The Collapse of Evolution*, Baker, p. 110
42. Parker, Gary, *What Is Creation Science?*, Master, p. 135-36
43. Baker, Sylvia, *Bone of Contention*, Creation Science Foundation, Ltd., p. 13
44. Meldau, F.J., *Why We Believe In Creation Not In Evolution*, Christian Victory Publishing, p. 150-51
45. Morris, H.M. and J.D. Morris, *Science and Creation*, Master, p. 65-66
46. Gish, D.T., *Evolution: The Fossils Still Say No!*, ICR, p. 149
47. Huse, S.M., *The Collapse of Evolution*, Baker, p. 109
48. Gish, D.T., *Evolution: The Fossils Still Say No!*, ICR, p. 155-60
49. Gish, D.T., *Evolution: The Fossils Still Say No!*, ICR, p. 176-77
50. Huse, S.M., *The Collapse of Evolution*, Baker, p. 104
51. Morris, H.M. and J.D. Morris, *Science and Creation*, Master, p. 73-75
52. Whitcomb, J.C. and H.M. Morris, *The Genesis Flood*, Presbyterian and Reformed, p. 285
53. Morris, H.M. and J.D. Morris, *Science and Creation*, Master, p. 76-77
54. Gish, D.T., *Evolution: The Fossils Still Say No!*, ICR, p. 198-200

Literature Citations: Chapter 8

1. Huse, S.M., *The Collapse of Evolution,* Baker, p. 102.
2. Lubenow, M.L., *Bones of Contention*, Baker, p. 45
3. Gish, D.T., *Evolution: The Fossils Still Say No!*, ICR, p. 241-42.
4. Lubenow, M.L., *Bones of Contention*, Baker, 1992, p. 48
5. Huse, S.M., *The Collapse of Evolution*, Baker, p. 143
6. Goodwin, G.G., *The Illustrated Encyclopedia of Animal Life*, Greystone Press, p. 197
7. Gish, D.T., *Evolution: The Fossils Still Say No!*, ICR, p. 233
8. Lubenow, M.L., *Bones of Contention*, Baker, p. 49-50
9. Gish, D,T., *Evolution: The Fossils Still Say No!*, ICR, p. 259
10. Lubenow, M.L., *Bones of Contention*, Baker, p. 167
11. Gish, D.T., *Evolution: The Fossils Still Say No!*, ICR, p. 259
12. Gish, D.T., *Evolution: The Fossils Still Say No!*, ICR, p. 234
13. Morris, J.D. and H.M. Morris, *Science and Creation*, Master, p. 85

14. Gish, D.T., *Evolution: The Fossils Still Say No!*, ICR, p. 234
15. Johnson, P.E., *Darwin on Trial*, Regnery Gateway, p. 80
16. Lubenow, M.L., *Bones of Contention*, Baker, p. 158
17. Gish, D.T., *Evolution: The Fossils Still Say No!*, ICR, p. 263-265
18. Lubenow, M.L., *Bones of Contention*, Baker, p. 162, 164
19. Gish, D.T., *Evolution: The Fossils Still Say No!*, ICR, p. 300
20. Lubenow, M.L., *Bones of Contention*, Baker, p. 165
21. Johnson, P.E., *Darwin on Trial*, Regnery Gateway, p. 80
22. Lubenow, M. L.. *Bones of Contention*, Baker, p. 124-125
23. Lubenow, M. L., *Bones of Contention*, Baker, p. 132
24. Lubenow, M.L., *Bones of Contention*, Baker, p. 120
25. Gish, D.T., *Evolution: The Fossils Still Say No!*, ICR, p. 280-81
26. Lubenow, M. L., *Bones of Contention*, Baker, p. 87
27. Baker, Sylvia, *Bone of Contention*, Evangelical Press, p. 14
28. Lubenow, M. L., *Bones of Contention*, Baker, p. 102-103
29. Lubenow, M.L., *Bones of Contention*, Baker, p. 89-90
30. Gish, D.T., *Evolution: The Fossils Still Say No!*, ICR, p. 281-282
31. Lubenow, M. L., *Bones of Contention*, Baker, p. 93
32. Morris, H.M., *The Long War Against God*, Baker, p. 199-200
33. Lubenow, M.L., *Bones of Contention*, Baker, p. 94
34. Morris, J. D. and H.M. Morris, *Science and Creation*, Master, p. 90
35. Morris, H.M., *The Biblical Basis of Modern Science*, Baker, p. 406
36. Meldau, F. J., *Why We Believe in Creation Not Evolution*, Christian Victory Publishing, p. 245
37. *Encyclopedia Britannica*, Encyclopaedia Britannica, Inc., Vol. 28:72
38. Lubenow, M.L., *Bones of Contention*, Baker, p. 137
39. Lubenow, M.L., *Bones of Contention*, Baker, p. 78-80
40. Lubenow, M.L., *Bones of Contention*, Baker, p. 84-85
41. Weaver, K.F., "The Search for Our Ancestors," *National Geographic*, Nov. 1985, National Geographic Society, p. 584
42. Lubenow, M.L., *Bones of Contention*, Baker, p. 59
43. Parker, G.E., *What Is Creation Science?*, Master, p. 152
44. Lubenow, M.L., *Bones of Contention*, Baker, p. 61-65
45. Morris, J.D., *Scopes: Creation on Trial*, ICR, p. 38
46. Lubenow, M.L., *Bones of Contention*, Baker, p. 61-65
47. Morris, J.D., *Scopes, Creation on Trial*, ICR, p. 38
48. Famighetti, Robert, editor, *World Almanac and Book of Facts*, K-111 Reference Corp., p. 853
49. Whitney, D.C., editor, *Reader's Digest 1983 Almanac and Yearbook*, The Reader's Digest Association, Inc., p. 849
50. Lubenow, M.L., *Bones of Contention*, Baker, p. 169
51. Johnson, P.E., *Darwin on Trial*, Regnery Gateway, p. 80
52. Gish, D.T., *Evolution: The Fossils Still Say No!*, ICR, p. 328
53. Lubenow, M.L., *Bones of Contention*, Baker, p. 42-43
54. Morris, J.D., *Scopes: Creation on Trial*, ICR, p. 40
55. Meldau, F.J., *Why We Believe In Creation Not In Evolution*, Christian Victory Publishing, p. 321
56. Morris, J.D., *Scopes: Creation on Trial*, ICR, p. 40
57. Barber, Raymond, *Sword of the Lord*, 22 August 1997, p. 6
58. Morris, J.D. and H.M. Morris, *Science and Creation*, Master, p. 81-82

59. Lubenow, M.L., *Bones of Contention*, Baker, p. 21
60. Lubenow, M.L., *Bones of Contention*, Baker, p. 83
61. Huse, S.M., *The Collapse of Evolution*, Baker, p. 37

Literature Citations: Chapter 9

1. Morris, H.M. and J.D. Morris, *Science and Creation*, Master, p. 113
2. Whitcomb, J.C. and H.M. Morris, *The Genesis Flood,* Presbyterian and Reformed, p. 176-79
3. Morris, H.M. and J.D. Morris, *Science and Creation*, Master, p. 80
4. Thomson, K.S., *Living Fossil*, W.W. Norton and Co., p.13-15
5. Thomson, K.S., *Living Fossil*, W.W.Norton and Co., p. 72
6. Morris, H.M. and J.D. Morris, *Science and Creation*, Master, p. 114
7. Morris, H.M. and J.D. Morris, *Science and Creation*, Master, p. 119
8. Morris, H.M. and J.D. Morris, *Science and Creation*, Master, p. 115
9. Morris, H.M., ICR Impact #241, ICR
10. Morris, H.M., *The Biblical Basis of Modern Science*, Baker, p. 354
11. Mackal, R.P., *Searching for Hidden Animals*, Doubleday, p. 75
12. Mackal, R.P., *Searching for Hidden Animals*, Doubleday, p. 63-66
13. Gish, D.T., *Dinosaurs by Design*, Master, p. 17
14. Compton's Pictured Encyclopedia, Vol. 3, F.E. Compton and Co., p. 510
15. Clark, Jerome, *Unexplained!*, 1993, Visible Ink Press, Detroit, p. 97-98
16. Heuvelmans, Bernard, *On the Track of Unknown Animals*, Kegan Paul International/Columbia University, p. 554
17. Heuvelmans, Bernard, *On the Track of Unknown Animals*, Kegan Paul International/Columbia University, p. 567
18. Heuvelmans, Bernard, *On the Track of Unknown Animals*, Kegan Paul International/Columbia University, p. 569-572
19. Morris, H.M., *The Remarkable Record of Job*, Baker, p. 115-17
20. Gish, D.T., *Dinosaurs by Design*, Master, p. 80
21. Ham, Ken, Andrew Snelling, and Carl Wieland, *The Answers Book*, Master, p. 33
22. Morris, H.M. and J.D. Morris, *Science and Creation*, 1996, Master, p. 122
23. Gish, D.T., *Evolution: the Fossils Still Say No!*, ICR, p. 106
24. Gish, D.T., *Dinosaurs by Design*, Master, p. 86
25. Cohen, Daniel, *Mysteries af the World*, Doubleday, p. 71
26. Cohen, Daniel, *Mysteries of the World*, Doubleday, p.74-76
27. Gish, D.T., *Dinosaurs by Design*, Master, p. 17
28. Gish, D.T., *Dinosaurs by Design*, Master, p. 60
29. Clark, Jerome, *Unexplained!*, Visible Ink Press, p. 220-21
30. Clark, Jerome, *Unexplained!*, Visible Ink Press, p. 225-229
31. Gish, D.T., *Dinosaurs by Design*, Master, p. 16
32. Gish, D.T., *Dinosaurs by Design*, Master, p. 58
33. Heuvelmans, Bernard, *On the Track of Unknown Animals*, Kegan Paul International/Columbia Univ., p. 583-84
34. Heuvelmans, Bernard, *On theTrac k of Unknown Animals*, Kegan Paul Intemational/Columbia Univ., p. 589-93
35. Mackal, R.P., *Searching for Hidden Animals*, Doubleday, p. 54
36. Ham, Ken, Andrew Snelling, and Carl Wieland, *The Answers Book*, Master, p. 22

37. Gish, D.T., *Dinosaurs by Design*, Master, p. 76
38. Ham, Ken, Andrew Snelling, and Carl Wieland, *The Answers Book*, Master, p. 25-29

Literature Citations: Chapter 10

1. Chittick, D.E., *The Controversy*, Multnomah, p. 229
2. Morris, H.M., *The Long War Arainst God*, Baker, p. 10
3. Whitney, D.J., *The Face of the Deep*, Vantage, p. 6
4. Chittick, D.E., *The Controversy*, Multnomah, p. .230
5. Morris, H.M., *Science and the Bible*, Moody, p. 89
6. Chittick, D.E., *The Controversy*, Multnomah, p. 235
7. Baker, S., *Bone of Contention*, ICR, p. 23
8. Chittick, D.E., *The Controversy*, Multnomah, p. 236
9. Morris, J.D., "Back To Genesis," August 1997, ICR, p.
10. Huse, S.M., *The Collapse of Evolution*, Baker, p. 26
11. Morris, J.D., "Back To Genesis," August 1997, ICR, p.
12. Baker, S., *Bone of Contention*, Creation Science Foundation, p. 23
13. Whitcomb, J.C., and H.M. Morris, *The Genesis Flood*, Presbyterian and Reformed, p. 334-37
14. Aardsma, G.E., "Myths Regarding Radiocarbon Dating," ICR, Impact No. 189,
15. Morris, H.M. and J.D. Morris, *Science and Creation*, Master, p. 320
16. Morris, H.M., *Science and the Bible*, 1986, Moody, p. 90
17. Huse, S.M., *The Collapse of Evolution*, Baker, p. 20
18. Morris, H.M., and J.D. Morris, *Science and Creation*, Master, p. 323
19. Morris, H.M. , *Biblical Basis of Modern Science*, Baker, p. 263
20. Morris, H.M., and J.D. Morris, *Science and Creation*, Master, p. 328-30
21. Whitney, D.J., *The Face of the Deep*, Vantage, p. 29-30
22. Whitney, D.J., *The Face of the Deep*, Vantage, p. 31-33
23. Morris, J.D., *The Young Earth*, Master, p. 70-71
24. Whitcomb, J.C., and H.M. Morris, *The Genesis Flood*, Presbyterian and Reformed, p. 379
25. Huse, S.M., *The Collapse of Evolution*, Baker, p. 23
26. Morris, H.M., *Biblical Basis of Modern Science*, Baker, p. 321
27. Morris, H.M., *Science and the Bible*, Moody, p. 80
28. Vardiman, L., *Sea-Floor Sediments and the Age of the Earth*, ICR, p. 5
29. Morris, J.D., *The Young Earth*, Master, p. 90
30. Vardiman, L., *Sea-Floor Sediments and the Age of the Earth*, ICR, p. 3-4
31. Vardiman, L., *Sea-Floor Sediments and the Age of the Earth*, ICR, ICR, p. 10
32. Vardiman, L., *Sea-Floor Sediments and the Age of the Earth*, ICR, ICR, p. 57
33. Morris, J.D., *The Young Earth*, Master, p. 97
34. Whitcomb, J.C., and H.M. Morris, *The Genesis Flood*, Presbyterian and Reformed, p. 164
35. Morris, J.D., *The Young Earth*, Master, p. 97
36. Whitcomb, J.C., *The World That Perished.*, Baker, p. 124
37. Huse, S.M., *The Collapse of Evolution*, Baker, p. 24
38. Morris, H.M. and J.D. Morris, *Science and Creation*, Master, p. 267
39. Whitcomb, J.C., and H.M. Morris, *The Genesis Flood*, Presbyterian and Reformed, p. 432

40. Morris, J.D., *The Young Earth*, 1994, Master, p. 102
41. Whitcomb, J.C., and H.M. Morris, *The Genesis Flood*, Presbyterian and Reformed, p. 162-64
42. Ham, K., A. Snelling, and C. Wieland, *The Answers Book*, Master, p. 77-79
43. Whitcomb, J.C. and H.M. Morris, *The Genesis Flood*, Presbyterian and Reformed, p. 297
44. Whitcomb and Morris, *The Genesis Flood*, Presbyterian and Reformed, p. 247
45. Whitcomb and Morris, *The Genesis Flood,* Presbyterian and Reformed, p. 292
46. Velikovsky, I., *Earth In Upheaval*, Doubleday, p. 162
47. Whitcomb, J.C., *The World That Perished*, Baker, p. 86
48. Vardiman, L., *Ice Cores and the Age of the Earth*, ICR, p. 1-4
49. Vardiman, L., *Ice Cores and the Age of the Earth*, ICR, p. 28
50. Vardiman, L., *Ice Cores and the Age of the Earth,* ICR, p. 4-6
51. Whitney, D.J., *The Face of the Deep*, Vantage, p. 102
52. Sagan, C., *Cosmos*, Ballantine, p. 63
53. Sagan, C., *Cosmos*, Ballantine, p. 67-68
54. Gore, R., "Halley's Comet," National Geographic, Dec. 1986, p. 761
55. Huse, S.M., *The Collapse of Evolution,* Baker, p. 28
56. Whitcomb and Morris, *The Genesis Flood*, Presbyterian and Reformed, p. 382
57. Sagan, C., *Cosmos*, Ballentine, p. 63
58. Morris, H.M., and J.D. Morris, *Science and Creation*, Master, p. 327
59. Davies, K., *Evidences For A Young Earth*, ICR, Impact No. 276
60. Wilson, W., "Science and the Bible," Sword of the Lord, 22 Aug. 1997, p. 14
61. Sagan, C., *Cosmos*, Ballantine, p. 186-187
62. Morris, H.M., *Biblical Basis of Modern Science*, Baker, p. 164
63. Barnes, T.G., "Young Age For the Moon and Earth," ICR
64. DeYoung, D.B., *Astronomy and the Bible*, 1989, Baker, p. 40
65. Huse, S.M., *The Collapse of Evolution*, Baker, p. 30
66. Ham, K., A. Snelling, and C. Wieland, *The Answers Book,* Master, p. 185-187
67. Humphreys, D.R., *Starlight and Time*, Master, p. 21-22
68. DeYoung, D.B., *Astronomy and the Bible*, Baker, p. 86
69. Morris, H. M., *Biblical Basis of Modern Science*, Baker, p. 174
70. Humphreys, D.R., *Starlight and Time*, Master, p. 11
71. Ham, Snelling, and Wieland, *The Answers Book*, Master, p. 189-192

Literature Citations: Chapter 11

1. Hick, John. *Encyclopedia Britannica*, Encyclopedia Britannica Inc. Vol. 16, p. 324
2. McDowell, Josh, *Evidence That Demands A Verdict,* Vol. 1, Here's Life Publishers, Inc., p. 63
3. Morris, H.M., *The Long War Against God*, Baker, p. 210
4. Rice, J.R., *In the Beginning*, Sword of the Lord, p. 48-53
5. Morris, H.M., *The Genesis Record*, Baker, p. 54
6. Whitcomb, J.C., *Early Earth*, Baker, p. 141
7. Morris, H.M. and J.D. Morris, *Scripture and Creation*, Master, p. 54
8. Rice, J.R., *In the Beginning*, Sword of the Lord, p. 37

9. Whitcomb, J.C., *Early Earth*,Baker, p.146
10. Rice, J.R., *In the Beginning*, Sword of the Lord, p. 40
11. Ham, Ken, *The Lie: Evolution*, Master, p. 113
12. Rice, J.R., *In the Beginning*, Sword of the Lord, p. 37-41
13. Whitcomb, J.C., *Early Earth*, Baker, p. 151-53
14. Whitcomb, J.C., *Early Earth*, Baker, p. 147-48
15. Morris, H.M. and J.D. Morris, *Scripture and Creation*, Master, 61-62
16. Hyles, J.F., "Logic Must Prove The King James Bible," (audio cassette), Hyles Publications, 8 April 1984
17. Hutson, Curtis, *Who Is A Fundamentalist?*, Sword of the Lord, 1982, p. 13
18. Rice, J.R., *God's Authority*, Sword of the Lord, p. 19-20
19. Ham, Ken, *The Lie: Evolution*, Master, p. 41-44
20. Morris, H. M., *Twilight of Evolution*, Baker, p. 23
21. Pyle, Hugh, *The New Modernism,* Sword of the Lord, p. 4
22. Hyles, J.F., "Logic Must Prove the King James Bible," (audio cassette)
23. Rice, J.R., *Our God-Breathed Book — the Bible,* Sword of the Lord, p. 31
24. Bryan, W.J., *The Bible or Evolution?*, Sword of the Lord, p. 29

Literature Citations: Chapter 12

1. Morris, H.M., *Long War Against God*, Baker, p. 53-54
2. Morris, H.M. and John Morris, *Society and Creation*, Master, p. 34-40
3. Whitcomb, John, *The Early Earth*, Baker, p. 63
4. Morris, H.M. and John Morris, *Society and Creation*, Master, p. 40-41
5. Morris, H.M., *Long War Against God*, Baker, p. 57-60
6. Morris Henry, ICR Impact No. 7, "Evolution and Modern Racism," p. 1
7. Morris Henry, ICR Impact No. 7, "Evolution and Modern Racism," p. 3
8. Morris, H.M., *Long War Against God*, Baker, p. 72
9. Humber, Paul, ICR Impact no. 227, "Evolution and the American Abortion Mentality," May 1992
10. Ham, Kenneth, *The Lie, Evolution*, Master, p. 85-86
11. Morris, H.M. and John Morris, *Society and Creation*, Master, p. 88-90
12. Ben-Law, David, *From Hitler's Hell to God's Peace*, Marshall Graphics, p. 14
13. Morris, H.M., *Long War Against God*, Baker, p. 78
14. Morris, H.M. and John Morris, *Society and Creation,* Master, p. 105-107
15. Morris, H.M., *Long War Against God*, Baker, p. 84
16. Morris, H.M., *Long War Against God*, Baker, p. 180-181
17. Humber, Paul, ICR Impact no. 172, *Stalin's Brutal Faith*
18. Morris, H.M., *Long War Against God*, Baker, pp 85
19. Smith, Shelton, editor, Sword of the Lord, April 1998, p, 2
20. Humber, Paul, ICR Impact no. 227, "Evolution and the American Abortion Mentality," May 1992
21. Morris, John, *Scopes: Creation on Trial*, ICR, p. 36
22. Morris, H.M. and John Morris, *Society and Creation* , Master, p. 63-65
23. National Right to Life Committee, *When Does Life Begin?*, 1993, p. 1-5
24. McLaughlin, Roy, Sword of the Lord, Jan. 26, 1996, p. 21
25. Humber, Paul, "Evolution and the American Abortion Mentality," ICR Impact No. 227, May 1992, p. 1
26. Morris, H.M. and John Morris, *Society and Creation*, Master, p. 18

27. Morris, H.M. and John Morris, *Society and Creation*, Master, p. 67
28. Morris, H.M. and John Morris, *Society and Creation,* Master, p. 60
29. Morris, H.M. and John Morris, *Society and Creation*, Master, p. 67-69
30. Wilson, Walter, Sword of the Lord, Aug. 22,1997, p. 14
31. Morris, H.M. and John Morris, *Society and Creation*, Master, p. 161
32. Moore, John, *Impact of Evolution on the Humanities and Science*, ICR Impact No. 53, November 1977, p. 3
33. Morris, H.M., *Long War Against God*, Baker, p. 48
34. Morris, H.M. and John Morris, *Society and Creation*, Master, p. 70
35. Ham, Kenneth, *The Lie: Evolution*, Master, p. 101
36. Morris, H.M., *History of Modern Creationism*, ICR, p. 18
37. Gish, D. T., *Creation Scientists Answer Their Critics*, ICR, p. 18
38. Morris, H.M. and John Morris, *Society and Creation*, Master, p. 167-169
39. Brock, Jim, Sword of the Lord, July 1, 1994, p. 16-17
40. Morris, H.M. and John Morris, *Sociey and Creation*, Master, p. 169
41. Morris, H.M., *History of Modern Creationism*, ICR, p. 70-73
42. Cornelius, R.M., *Scopes: Creation on Trial*, ICR, p. 2-3
43. Cornelius, R.M., *Scopes: Creation on Trial*, ICR, p. 20
44. Morris, H.M., *History of Modern Creationism*, ICR, p. 70-73
45. Cornelius, R.M., *Scopes: Creation on Trial*, ICR, p. 4
46. Morris, H.M., *History of Modern Creationism*, ICR, p. 70-72
47. Morris, John, *Scopes: Creation on Trial*, ICR, p. 29
48. Morris, H.M., *Long War Against God*, ICR, p. 121-122
49. Morris, H.M., *Long War Against God*, ICR, p. 126
50. Sexton, Clarence, *Unmasking the New Age Movement*, Sword of the Lord, p. 13
51. Ferguson, M., *The Aquarian Conspiracy*, Tarcher, p. 158-159
52. Ferguson, M., *The Aquarian Conspiracy*, Tarcher, p. 183
53. Marrs, T., *Dark Secrets of the New Age*, Crossway, p. 125-126